THE
IDEA
MACHINE

How Books Built Our World
and Shape Our Future

Joel J. Miller

Prometheus Books

Essex, Connecticut

Prometheus Books

An imprint of The Globe Pequot Publishing Group, Inc.
64 South Main Street
Essex, CT 06426
www.globepequot.com

British Library Cataloguing in Publication Information available

Library of Congress Cataloging-in-Publication Data available

ISBN 9781493088935 (cloth)
ISBN 9781493088942 (electronic)

To Naomi, our littlest reader

CONTENTS

The most technologically efficient machine that man has ever invented is the book.

—Northrop Frye

Without books, history is silent, literature dumb, science crippled, thought and speculation at a standstill. Without books, the development of civilization would have been impossible.

—Barbara Tuchman

INTRODUCTION

The Forgotten Technology

WHEN *MIT TECHNOLOGY REVIEW* ASKED MICROSOFT COFOUNDER BILL Gates to curate a list of recent technological breakthroughs, the magazine also featured a short piece he had prepared on the books that informed his selections. "Whenever I want to understand something better," he said, "I pick up a book. Reading is my favorite way to learn about a new subject— whether it's global health, quantum computing, or world history."

Gates is famous for toting around a bag full of books, packed each day by his executive assistant. He's also famous for his annual "think weeks" in which he sequesters himself from the world, drinks gallons of Diet Coke, and reads even more.

For the assignment, Gates identified developments in artificial intelligence, robotics, medicine, energy, food, and even sanitation. Had the picks gone beyond MIT's narrow timeframe, however, I imagine he would have included the book itself. In fact, it's safe to say none of these breakthroughs would have occurred at all had humans failed to first invent the book.

The book, as I argue in the pages ahead, is one of the most important but overlooked factors in the making of the modern world. Why this lack of appreciation—or even awareness? Arguably, the book is a victim of its own success. Familiarity usually breeds more neglect than contempt. We fail to recognize the book for what it is: a remarkably potent information technology, an idea machine.

One well-known story might help reacquaint us with its powers and launch us into the subject.

Augustine's Astounding Mobile Device

In a pivotal scene of his *Confessions*, Augustine retreats to a garden. Distraught, he clutches a book as he goes. Thinking to comfort him, Augustine's friend Alypius follows him outdoors. The pair finds a place to sit far from the house they're sharing, but Augustine is inconsolable. Not wanting Alypius to see his tears, he wanders off once more, leaving his friend and his book on a bench. Finally collapsing in the shade of a nearby fig tree, Augustine sobs.

At that moment the singsong voice of a child wafts over the garden wall. "Tolle lege," he hears. "Take and read." He first assumes it's just the playful call-and-response of a game. But then he remembers his book! He takes it as a sign. Running back, he grabs the volume, opens it at random, and falls upon a passage that immediately salves his troubled spirit.

Many of us have had similar moments. A chance turn of pages produces a line that stays with us for days, maybe decades. That's Augustine's emphasis: the transformation wrought by words encountered. Closing the book, Augustine is free to muse on what he has read. What was inside the book is now inside Augustine. He can reflect on it, try it on, fit it into his life, and fit his life around it.

Books and their effects play a recurring role in the *Confessions*. As a boy, Augustine is moved by the death of Dido in Vergil's *Aeneid*. He's shaped by reading books by Cicero, astrologers, Manichaeans, and Platonists. Just hearing a visitor discuss *The Life of St. Anthony*, a popular biography of a monastic pioneer, triggers Augustine's crisis in the garden. But there's much more going on in this brief scene.

Augustine's book was a collection of St. Paul's letters. The garden revelation happened in the year 386 CE—by which time the apostle had been dead more than 320 years. "The writer has some privileges," says the Russian novelist Eugene Vodolazkin. "One of them is that he can continue the conversation with the reader after death." He says in another place, "It's so strange: the person is already gone, but, yes, a book continues to live." It is strange, though we've grown inured to it.

Augustine, distraught, takes and reads. Painting by Benozzo Gozzoli. *SOURCE:* HEMIS, ALAMY.

2

Books present a curiously casual form of necromancy. Standing in my home library, I see the names of people whose voices echo decades, centuries, millennia after their final repose. To reanimate their thoughts and stories, I need only do like Augustine: take and read. Conversations begun in one age suddenly resume in another, each encounter forging a new link in a long "human chain of paper and flesh," to use a phrase from another novelist, Rodrigo Fresán.

And notice that Paul does more than speak to Augustine. He also speaks to Alypius, who picks up the reading where Augustine left off. How? When he finishes, Augustine says, "I placed my finger or some other marker in the book and closed it." It's a small but revealing detail. What Augustine finds by chance, Alypius now accesses by intention.

By marking the page, they fix the finding and enable future reference. Now either or both of them can return, reengage, and reappraise. Multiply the bookmarks, and they can balance one idea against others in the same or additional books. They can then use this latticework of ideas to support new insights, which can be captured in still more books and shared with subsequent readers, each of whom can have similar experiences.

Then there's portability. Augustine carries his copy where he wills and shares it with whom he desires. Books enable us to grasp whole worlds and shuttle them where we like. We can spread them across tables, move them indoors and out, arrange them on shelves to suit our interests and amusements, or even surprise ourselves with chance encounters.

We can, for example, place books in conversation whose authors never would or could have talked in life. Sidonius Apollinaris, an aristocrat and bishop born at the end of Augustine's life, described visiting a large library on a country estate in Gaul where he lived. Pagan and Christian authors sharing "a similarity of style" were shelved side by side "though their doctrines [were] different." The owner paired pagan poet Horace with Christian poet Prudentius and placed his Augustine next to his Varro—entire centuries being the smallest of their differences.

Books can accomplish this sidling and sparring inside their covers too, creating emergent notions and interpretive layers that expand and transform their contents. To say it again, Augustine read a *collection* of St. Paul's letters. Linger on what that can mean. Paul never wrote a book, as such. Rather, he addressed ad hoc letters over many years involving different audiences and concerns. These texts were later gathered and bound in a single volume. Books do for ideas what mosaics do for stones. Making a book of Paul's letters was a creative act that transformed how they were read and understood. The same can be said about the entire Bible, a book made of other books and bits. Taken as a whole, it yields pictures unseen in its individual parts.

That idea takes us to possession itself. Writers have only limited claims on their words once published. Readers use them as they will. How many have sailed past Augustine's salient passage without a second's thought? St. Paul in one person's hands is different from St. Paul in another's; just ask the Protestant Reformers. This is Augustine's Paul, his copy to read, to understand, to internalize, to utilize as he desires.

Books permit access to (and acquisition of) ideas that once lived in the minds of others. Once apprehended—accurately or not—they can be put to uses unimagined by their originators. Every discipline from medicine to theology, mathematics, politics, philosophy, and science provides examples. Isaac Newton didn't stand on the shoulders of giants, as he said; he clambered atop their books.

Augustine tells the story in the garden because of what the content of his book did for him. In that sense, he and Gates share a similar appreciation for the book. As a bearer of ideas alone, the book deserves our admiration. But reflecting on these additional facets of Augustine's experience—which the *form* of the book made possible—we see that the book's technological features and potentialities are equally important. Beyond what value the subject possesses, the object brings it forward and enables much else besides.

"Books are a uniquely portable magic," novelist Stephen King once said. It's an observation I take as confirmation of another, known as Clarke's Third Law: "Any sufficiently advanced technology is indistinguishable from magic." We tend to equate books with mere information. In doing so, we miss that they are an information technology. And not just any information technology—no, in understanding how our world came to be, books emerge as an essential technology. Books are hardware as well as software, and it's the combination that explains their peculiar power and effect.

The Product of Books

My enchantment began in childhood. Dad was an English teacher, and Mom loved mysteries. Napoleon and Snowball, Father Brown, Atticus Finch, Poirot, Romeo and Juliet, and Miss Marple all vied for attention on shelves throughout the house, further crowded by memoirs, biographies, religion, romances, spy thrillers, economics, history, philosophy, comics, politics, cookbooks, and at least two Volkswagen technical manuals.

I was a voracious if uneven reader as a child. Fantasy and survival tales captivated me. I devoured the Tarzan novels and loved mythology, particularly the Irish stuff. I discovered early the treacherous chasm between books and movies, not always to my delight. When I reached the last page of *Swiss Family Robinson* having encountered no pirates, I felt robbed. And Rambo actually—spoiler alert!—dies in *First Blood*; author David Morrell resurrected him for a sequel when filmmakers decided to let the traumatized Vietnam vet live. But the occasional curveball only heightened my interest. That the same stories could differ from one teller to the next fascinated me.

Then, in my teens, I started writing. I was around nineteen or twenty when I received a gracious rejection from the celebrated editor and publisher Morgan Entrekin for a cringe-worthy attempt at political satire I had submitted. He spared me inevitable humiliation and the world involuntary eye rolling.

Though I kept writing and eventually published a few books, I also stayed close to literature by other means. I got my first real job at the Almost Perfect Bookstore in Roseville, California, a used bookshop that has since closed its doors. I also sorted, shelved, and sold books at Borders Books & Music, now gone as well. Best of all, I landed an editorial gig at Thomas Nelson Publishers, in time making my way to vice president and publisher of a division.

My interest in book history started in those days as an editor and publisher. I was like a chef investigating the cuisine he prepared. I started reading the memoirs and observations of editors and publishers—people like Maxwell Perkins, William Jovanovich, Henry Regnery, and André Schiffrin. That led to the history of publishing and then to the history of books themselves. I became fascinated by the development of their use and how, in turn, their use featured in human cultural development.

This book isn't about me, but I share my background to shed light on my interest and perspective. I have read books, written books, peddled books, edited books, commissioned books, and published books. Whether personally or professionally, my world would not be the same without them. And, as I argue in the pages ahead, neither would *our* world.

Considering their impact, books have played a distinctly undervalued role in making us who we are. "I am a product of . . . endless books," said C. S. Lewis in his memoir, *Surprised by Joy*. It's true for us all, whether we realize it or not. Which prompts a simple but possibly vexing question.

What Is a Book, Exactly?

It's worth answering at the outset. Most would say sheets of paper sandwiched between covers and bound on one side with glue or thread—or maybe digital files purchased with a click and consumed via mobile device. That's not wrong. Nor is it the full story.

The first books were soft clay tablets gouged with reeds and then dried in the sun; cuneiform takes its name from the Latin for all those little wedge-shaped characters that look like bird tracks in the sand. Books have since taken many other forms as well: scrolls, waxen tablets, and codices of papyrus, wood, and parchment. Nor do these exhaust the materials and formats.

So, how do we define such an extraordinary tool given the range of shapes, substances, and uses? Faced with entries for "books" in wills and bequests, Roman lawyers answered that question ages ago. "When books are bequeathed," said one jurist, "not merely rolls of papyrus, but also any kind of writing which is contained in anything, is understood." Easy enough. There's more, as we'll see, but for now let me venture the following definition based on what's above and what's to come:

The book is a portable collection of written ideas, designed to elevate the human mind beyond its natural limits of experience, memory, distance, and time; it's a vessel for numbers, narratives, laws, and lyrics; it facilitates history, politics, philosophy, religion, science, and self-

Sidonius Apollinaris, studiously at work in his library. Engraving by André Thevet.
SOURCE: AUTHOR'S COLLECTION.

discovery; it enshrines traditions while providing direction as they shift and grow; it informs the ignorant, reminds the learned, travels far, and cheats death.

Tools transform their users. As their use spread, books altered human thinking, changed how we organized information and people, and helped us reimagine everything from government to education, the sexes, worship, the natural world, ourselves. Muhammad called Jews and Christians people of the book. But the truth is, whether directly or indirectly, religious or not, that name fits us all.

Books enable us to stack up ideas and see whether they can bear the weight of new associations and connections. Even a small library enables us to trade one vista for another, layer those vistas atop each other in novel combinations, or imagine altogether new landscapes of our own. Books thus broaden and enrich us, and we are heir to all who have in the past been so broadened and enriched.

Scholars have sometimes focused on ideological answers for how we became modern. Spotting flaws in this approach, others have stressed material causes: economics, tools, even diseases. (Compare, for example, Richard Weaver's *Ideas Have Consequences* with Jared Diamond's *Guns, Germs, and Steel*.) I say it's both, and from where I stand, the credit goes to books—particularly the interplay of content, format, and use—for creating ways we thought and believed in the past and think and believe today.

The Pages Ahead

How did books propel such an evolution? In the pages ahead, I lay out my answer by looking at various features and uses of the book through time. If we think of the book as a tool, as an information technology, we must ask what it has enabled us to accomplish and how. The path to now is paved with books. To discover how, you need only follow Augustine: take and read.

We'll start in Greece during the fifth and fourth centuries before the Common Era, when books began edging their way into Athenian culture thanks to an uptick in papyrus imports. Supply increased, prices fell, and use spread. Books were present before, albeit mostly scrawled on animal hide. Papyrus changed the game.

By the time Herodotus was writing his *Histories* (between 426 and 415 BCE), he could say only foreigners still wrote on goat or lamb hides, as the Greeks used to do. Less than one hundred years later, the orator Demosthenes said a small sheet of papyrus could be purchased for a couple copper coins. Papyrus wasn't exactly cheap, but the wealthy had no trouble affording it.

As with any emerging technology, the new tool for capturing thoughts attracted both praise and criticism. Even when readers were few in number, books began causing a stir.

Marginalia: Ideas across Time

Picture a simple, two-dimensional grid. We can plot an idea's expression along the x-axis and its specificity along the y-axis. As you sharpen and define the idea, it moves up the y-axis. As you articulate and share it, it moves rightward along the x-axis. When an idea moves far along both axes, we can say it's widely understood.

Most ideas never make it that far. Think, for instance, about a hunch. It lingers at the bottom of the y-axis—a shadowy, gooey notion in the mind. And since it's not yet clear enough to talk about, it also remains low on the x-axis. Ideas that cluster in this quadrant of the grid are *vague and private*.

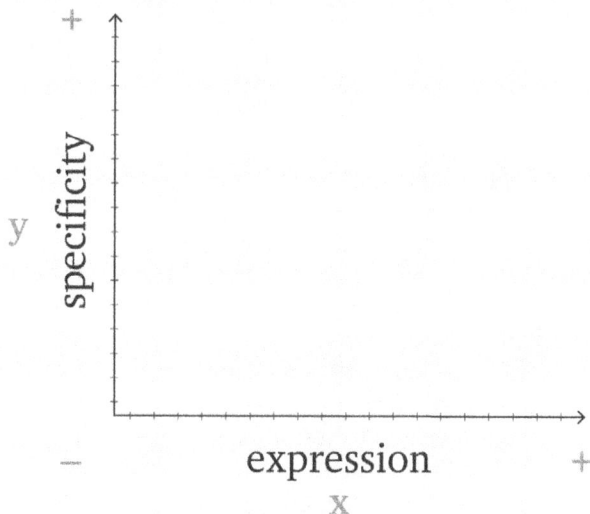

Plotting expression and specificity on x- and y-axes.
Illustration by Mike Burns. *SOURCE:* AUTHOR'S COLLECTION.

We might also have precise, fully formed ideas but leave them unsaid. Perhaps they're too personal, complicated, or simply ill timed to express. These would be high on the y-axis, low on the x-axis. Ideas that cluster in this portion of the grid are *precise but private*.

Of course, we sometimes blurt, post, text, email, or otherwise broadcast half-formed ideas. Ideas in this low-y, high-x cluster are *vague but public*. Only when ideas are both fully formed and fully expressed do they cluster in the high-*y*, high-*x*, *precise-and-public* quadrant.

Charles Darwin's theory of evolution follows this trajectory. It started as a hazy intuition, sparked by his observations during the voyage of the HMS *Beagle*. He noticed differences among species that ignited his curiosity about natural adaptation. At that stage there was no theory, only vague impressions and questions. But over time Darwin's ideas migrated up the *y*-axis, taking shape and gaining clarity. A theory began to emerge, a detailed understanding of natural selection. It became *precise but private*, recorded in notebooks and shared only with a trusted few. The leap to *precise and public* came with the publication of *On the Origin of Species*, articulating his once-private thoughts for a wider audience.

The book turned a hunch into a revolutionary idea, triggering immediate debate and forever changing the study of biological science and other fields. Isaac Newton's *Principia Mathematica* traveled a similar route to similar effect. So, too, for that matter, did Harriet Beecher Stowe's *Uncle Tom's Cabin*.

Stowe's outrage over the injustice of slavery simmered in private for years. Eventually, her general impressions took shape through writing, turning into characters and plot lines that illuminated the brutality of slavery. Stowe's conviction gained specificity as a private thought, but it became a social force only when she shoved it into the public sphere. Once published, *Uncle Tom's Cabin* changed the terms of the debate, shaping attitudes and becoming a rallying cry for abolition. By moving across the idea grid from *vague and private* to *precise and public*, Stowe's convictions changed the course of American history.

But here's the twist: Two dimensions aren't enough. Ideas don't simply become more precise or more public; they can persist, resonate, and evolve. To capture this dynamic, we need a *z*-axis, representing time itself. Now the idea grid extends into a third dimension, where ideas linger. And, as a technology, it's books that make this *z*-axis possible. By preserving ideas in a physical form, books allow thoughts to reach beyond their original moment, crossing decades, centuries, even millennia.

The *z*-axis lets ideas outlive their creators. When Darwin wrote *On the Origin of Species*, he did more than make his thoughts precise and public; he ignited a conversation that would extend across generations. We can say the same about Newton's *Principia* and Stowe's *Uncle Tom's Cabin*. The *z*-axis doesn't simply preserve ideas; it can also ensure that they take root in our cultural consciousness.

That said, such persistence comes with complications. An idea that once seemed clear can become obscure, sliding down the *y*-axis as it gets

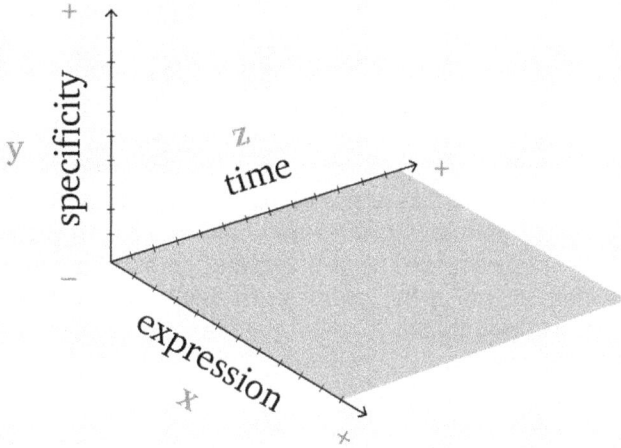

Plotting time along the z-axis. Illustration by Mike Burns.
SOURCE: AUTHOR'S COLLECTION.

refracted through the lenses of different eras and distant readers. What Darwin wrote in the nineteenth century possessed a particular clarity to his contemporaries, but as his ideas have traveled along the z-axis, they've encountered new interpretations, discoveries, and paradigms, resulting in a lower position on the x-axis as the *Origin* commands less direct public attention than it once did. And the same, of course, holds for the thinking of Newton and Stowe.

Long before Darwin, Newton, or Stowe, the philosopher Socrates understood this tension. He worried that books were too static and unable to adapt or respond to new questions. A book, he argued, cannot answer back when challenged; it says the same thing to everyone, regardless of their time. But that's only half the story. Books may be fixed, but their meanings are not. Each reader, in each era, brings a new perspective, reinterpreting and reshaping the ideas within. The z-axis is not just about persistence; it's also about the ongoing dialogue between past and present.

The z-axis doesn't simply let ideas endure; it makes them dynamic, allowing them to be recast, reimagined, and reconfigured by readers from one generation to the next. Books enable ideas to encounter new audiences, to be read with fresh eyes, to inspire new thoughts. A text's journey along the z-axis is full of reinterpretations, revisions, rebuttals, even misreadings. An idea that once seemed precise can become more ambiguous as its original context fades, only to find new relevance when that context shifts again.

This is the paradox of books: They freeze ideas in time and yet also make them endlessly revisitable and revisable. With the z-axis, the idea grid becomes more than just a map of how ideas grow. It's a map of how they live, how they reach across time, and how they remain part of our collective consciousness. The x-axis of expression and the y-axis of specificity show us how thoughts take shape, but the z-axis reveals how they endure, adapt, and transform. And books are the information technology that allows ideas to make this journey, inviting new interpretations and shaping the world we inhabit.

Ready to see how?

BASICS

Si nous assimilons le livre à une machine, c'est parce que ce mot évoque le concept de "mécanisme fonctionnel": agencement de pièces dont l'interaction tend a l'obtention d'un résultat délibérément poursuivi.

If we equate the book with a machine, it's because this word evokes the concept of a "functional mechanism": an arrangement of parts whose interaction tends to achieve a deliberately pursued result.

—Carla Bozzolo, Dominque Coq,
Denis Muzerelle, and Ezio Ornato

A book is a machine to think with.

—I. A. Richards

Chapter 1

SOCRATES, TECHNOPHOBE?

Hot Book Juice

In Aristophanes's comedy *The Frogs*, written in 405 BCE, the god Dionysus tires of contemporary playwrights. He longs for an earlier, more exciting generation of writers. So he does what any of us would probably do: He travels to Hades to retrieve the recently departed Euripides. Hilarity ensues as Dionysus buffoons his way through the realm of the dead, dressed cartoonishly like his half brother Heracles, who had once made the same trip.

As it happens, Hades houses many dead playwrights. Aeschylus, who had died fifty years before Euripides, reigns as the undisputed champion—until Euripides shows up, that is. This newcomer, having impressed the denizens of the dead, vies for Aeschylus's throne. Who is the greatest? The only solution is a duel, a battle of wits and words.

The two authors square off, trading lines and trash talk. Aeschylus represents an older sort, one unreliant on books in composing his stories. His opponent, by contrast, is a newfangled bookish writer. Aeschylus wastes no time in targeting this feature, tagging Euripides a "phrase-collector." The dig implies Euripides is unoriginal; he just recycles other people's ideas, gleaned from his reading.

Meanwhile, Euripides takes aim at Aeschylus's heavy verbosity, which Euripides says he can slim down with a diet of "beetroot, light phrases, little walks, hot book-juice, and cold reasoning." Another translation offers "chatter-juice strained off from books." And so they go, back and forth, criticizing their opponent's work and insulting each other.

Euripides, gripping a scroll while receiving a theater mask, with a statue of Dionysus looking on. Marble bas-relief. SOURCE: ARCHAEOLOGICAL MUSEUM OF ISTANBUL; JOHN-GRÉGOIRE, WIKIMEDIA COMMONS.

As the contest comes to a close, Aeschylus calls for a final judgment and takes another potshot at his foe's reliance on books:

> *Come, no more line-for-lines! Let him jump in*
> *And sit in the scale himself, with all his books.*

Euripides, in other words, doesn't measure up, even if he's granted the extra weight of his library. Indeed, Dionysus is compelled to judge in favor of Aeschylus. ("You traitor," says Euripides.) But the comedy also reflects some amusing ambivalence. There is, for instance, a lofty reference to the audience and their supposed facility with books and the written word:

> *Oft from long campaigns returning*
> *Thro' the devious roads of learning*
>   *These have wandered, books in hand;*

Nature gave them keen discerning
Eyes; and you have set them burning!
Sharpest thought or deepest yearning—
    Speak, and these will understand.

The audience's capacity for understanding is, Aristophanes says, reflected by their ability to read. He equates literacy with smarts. In a time of marginal literacy, the idea that all these theatergoers were capable of following along with the playbook or any other text is offered in jest, designed to flatter the listeners. Surely some in the crowd could read; the rest enjoyed being talked about as if they could.

Aristophanes's *The Frogs* illumines a moment when a new technology entered society. As with any new technology, there were early adopters, along with more reluctant users and those who eschewed the new.

Socrates's Complaint

The philosopher Socrates supposedly sneered at writing and books. One of his students, Xenophon, tells of a time when Socrates discussed the subject with Euthydemus, a young man who had amassed a significant library.

Reading books by popular poets and sophists has inflated Euthydemus's ego. At least that's how Xenophon presents it: "He held himself to be already superior to his contemporaries on account of wisdom and had great hopes of surpassing everyone in being able to speak and take action"—that is, using rhetoric for political purposes.

Euthydemus plans to collect as many books as he can. But Socrates is unimpressed. Book learning is far too narrow. Any other worthy art or activity requires a teacher. But somehow Euthydemus believes he can pick up everything he needs to manage the affairs of government from books alone.

To expose the error, Socrates begins poking around the edges of the young man's ambition. He starts by praising Euthydemus's desire for wisdom, evidenced by his book collecting. Physicians had written important books, says Socrates. Is Euthydemus planning on becoming a doctor? No.

Then an architect? Not that either. Maybe a mathematician? No again. An astronomer? Still no.

Failing those employments, says Socrates, Euthydemus must be studying to become a rhapsode, one of the traditional performers of Homer's poems, the *Iliad* and the *Odyssey*. "They say that you possess also all the verses of Homer," Socrates says. But Euthydemus likewise rejects this profession. He thinks too highly of himself; rhapsodes can memorize verses from books, he says, but otherwise aren't too smart.

No, what's important, the two men agree, is justice and the ability to determine it. Euthydemus is convinced that his book learning enables him to judge between justice and injustice. Socrates believes otherwise and proposes an ingenious test.

The cagey philosopher suggests arranging two columns, writing *J* at the top of one for justice and *I* at the top of the other for injustice. He next presents Euthydemus with several easy cases: lying, deception, mischief, and enslaving. All these Euthydemus correctly files under *I*. Well done! But then Socrates complicates matters by offering examples in which seeming injustices might actually be just—cases requiring true discernment. "What about if someone should," says Socrates, "steal or seize a friend's sword . . . when his friend is dispirited, out of fear he might destroy himself?"

By placing challenging cases such as this one before the boy, Socrates undermines Euthydemus's confidence. "I . . . no longer trust my answers," Euthydemus admits. By the end of the conversation, he realizes his book learning is mostly useless for the weightier questions of life. "I probably know nothing at all," he says.

Euthydemus is unready for politics. According to this view, books are no substitute for a teacher. Even trades that utilize books for instruction—architecture, astronomy, medicine, and others—have masters who train apprentices. The person who depends on books as their sole source of knowledge and learning is likely to be left with nothing to show for all their hours with their nose inches from a scroll.

Xenophon wrote his *Memorabilia* in 371 BCE, almost thirty years after Socrates's death in 399. But there's another account from roughly the same time that corroborates and even amplifies Socrates's apparent attitude. It's by another one of his students, the more famous Plato.

Tricky Invention

Plato's dialogue *Phaedrus* begins as Socrates chances upon a friend of that name strolling outside Athens late one morning. Phaedrus has just spent time listening to the famed orator Lysias present a speech about seduction.

Socrates feigns interest and offers to walk along if his friend will recite the speech. Phaedrus objects: There's no way he could recite the whole thing verbatim! Nonetheless, Socrates insists: Phaedrus enjoyed the speech, no? He must have nagged Lysias to read it, perhaps several times. And that's not all. "Eventually he [Phaedrus] borrowed the scroll himself," says Socrates, speaking of his friend in the third person, "and pored over those parts of the speech he particularly wanted to look at, and continued with this, sitting in his place from daybreak onwards, until he got tired and went for a walk, by which time . . . he knew the speech by heart, unless it was really quite long." All Phaedrus needed was to bump into some unlucky listener to try out his new learning!

Phaedrus ignores the accusation. True, he had borrowed the speech. But Phaedrus also insists he hasn't memorized it word for word. He caught enough to get the gist, which he can summarize for Socrates. But Socrates isn't satisfied.

"I suspect you've got the actual speech," says Socrates. "I have no intention of letting you practice on me when Lysias is here too." Sure enough, Phaedrus has the speech tucked in his sleeve. Lysias is a person back in Athens, but he's also a string of letters scrawled across a sheet of papyrus, rolled up and hauled around. Phaedrus could remember Lysias—reassemble his words by sheer mental effort—or he could simply refer to the text, revealing his friend bit by bit in narrow columns while unrolling the book.

Surely Phaedrus's memory is better than he lets on. But probably not that good. He doesn't push it with Socrates; instead, he offers to find a place to sit and read the speech.

Socrates and Phaedrus, discussing Lysias's speech. Illustration by Władysław Witwicki. *SOURCE*: WIKIMEDIA COMMONS.

The dialogue, considered one of Plato's best, presents more than playful banter. The speech and its contents are largely a pretext for discussing the methods and merits of rhetoric.

In the dialogue's most famous passage, Socrates tells the story of ibis-headed Theuth, an Egyptian god who invented not only dice and board games but also mathematics, geometry, and astronomy. Given that the masters of such disciplines prescribe books to students—as Socrates mentions in his conversation with Euthydemus—it's no surprise to find that Theuth also invented writing.

One day Theuth called on Thamus, the king of Egypt. He was eager to display his inventions and sell them to the king. He had hoped the ruler would encourage their use—and, we presume, increase Theuth's fans and admirers. But King Thamus wasn't so easily sold. He queried the god on each and every invention, weighing pros and cons and sharing his critique. How would the king come down on writing?

Theuth depicted with a stylus and ink palette. *SOURCE:* SABENA JANE BLACKBIRD, ALAMY.

"Your Highness," said Theuth, "this science will increase the intelligence of the people of Egypt and improve their memories. For this invention is a potion for memory and intelligence." The word for "potion" here is *pharmakon*, which can be translated as medicine or elixir ("hot book juice" again). But it can also indicate poison.

Thamus responds. "One person has the ability to bring branches of expertise into existence," he says, "another to assess the extent to which they will harm or benefit those who use them." Theuth only sees the upside of his creation, but Thamus fixates on the negative. "It will atrophy people's memories," he says, adding, "Trust in writing will make them remember things by relying on marks made by others, from outside themselves, not on their own inner resources, and so writing will make the things they have learnt disappear from their minds. Your invention is a potion for jogging memory, not for remembering."

This use of "memory" here refers to something other than a mere mental record. Thamus refers to deep, personally held knowledge, especially awareness of higher truths internalized and integrated through reincarnation over lifetimes or somehow acquired directly from the heavenlies. Memory, as Socrates says earlier, is where the philosopher accesses the divine.

At best, Thamus believes that books can remind readers of what they already know; at worst, readers might cram their head with so much superficial or worthless information that they lose hold of what really matters. Nor is that the only problem.

Not only is Theuth wrong about memory, according to Thamus, but writing also undercuts intelligence. "You provide your students with the appearance of intelligence, not real intelligence," he says. "Because your students will be widely read, though without any contact with a teacher, they will seem to be men of knowledge, when they will usually be ignorant." In other words, they will be like the bookish Euthydemus.

Unlike humble and repentant Euthydemus, however, Thamus imagines a false sense of knowledge will render these readers insufferable. "This spurious appearance of intelligence will make them difficult company," he says. No one likes a know-it-all, especially when that know-it-all actually knows nothing.

Books convey knowledge but don't always confer it, so it's easy for readers to assume they understand ideas better than they do—or pretend as much. And books can easily end up in the hands of unintended or unlearned readers who misunderstand and abuse them. "Once any account has been written down, you find it all over the place," says Socrates, "hobnobbing with completely inappropriate people no less than with those who understand it, and completely failing to know who it should and shouldn't talk to."

Thus books require support from vested authorities to ensure that they are properly understood. Something written, says Socrates, "always needs its father to come to its assistance." Historically, the heads of reading communities—priests, rabbis, professors, and the like—have guarded important books and disciplines to facilitate understanding and protect the books in their charge from misinterpretation (not that these figures always agree among themselves).

These figures provide one additional service: Books, as Socrates says, make poor conversation partners and debate opponents. If questioned, a speaker can restate, clarify, or bolster his case. A book can say only what it says, however insufficient to the moment. "They just go on and on," complains Socrates, "forever giving the same single piece of information." These guardians of the text also serve as interpreters and expositors—people who can apply a message beyond its original setting or circumstance.

Phaedrus is easily swayed by Socrates's story and arguments. Only a fool believes scribbling words in books preserves knowledge, and it takes the same sort of folly to assume that reading these books produces a useful understanding of that knowledge. "Written words can do [no] more than jog the memory of someone who already knows the topic that has been written about," says Socrates. "Quite so," concedes Phaedrus.

But not so fast.

Socrates, Hypocrite?

People have long taken these dialogues by Xenophon and Plato as evidence that Socrates saw little value in writing and books. But here's a curveball: Xenophon and Plato's works here demonstrate the value of writing and books. Despite the surface claims, there's plenty in both Xenophon's *Memorabilia* and Plato's *Phaedrus* that undercuts any strident adherence to their mentor's argument—not least the fact that they preserved his opinions in books and we're busy discussing them now. There's more.

Socrates says trusting books to deliver "clear and reliable" information is a fool's errand. But he uses books for just that purpose himself, admitting the use and—one suspects—joy of reading. Earlier in *Memorabilia*, for instance, Xenophon reports a conversation between his mentor and the sophist Antiphon. "Reading collectively with my friends," says Socrates, "I go through the treasures of the wise men of old which they wrote and left behind in their books; and if we see something good, we pick it out."

Not only did Socrates use books to jog the memory as he seems to disdain (and even refers to as an empty amusement), but his practice likewise betrays reliance on books for access to these thoughts in the first place. Had not these wise men composed and captured their thoughts in books, Socrates would be unable to consult them.

The fact that Socrates read these books with his friends is also interesting. While he listens to books being read aloud and may even do some of the reading himself, there's no mention of a teacher as he says is necessary. Perhaps he considers the community the teacher. Perhaps he's above such needs. That seems so in his reading of the philosopher Anaxagoras, which Plato describes in another dialogue, the similarly sounding *Phaedo*.

Socrates hears rumors that Anaxagoras believes Mind (Greek *Nous*) "orders and is the cause of all things." This news thrills Socrates to hear, so he races out and acquires Anaxagoras's books for himself. "I seized the books very eagerly, and read them as fast as I could," he says, before admitting disappointment. It turns out Anaxagoras taught nothing of the sort, focusing instead on material causes. "All my splendid hopes were dashed to the ground," says Socrates, "for as I went on reading I found that the writer made no use of Mind at all." As depicted here, Socrates requires nothing more than Anaxagoras's books to critique his thought.

Further, as the story of Euthydemus reveals, Socrates allows that books are useful to students of trades or disciplines that rely on the technology—the same sorts of disciplines Theuth supposedly invented along with writing. Though the books' value is fully realized only with the help of a teacher, the sciences and the arts require mastering information, formulas, data, processes, and methods that are best recorded in books for consultation by the student and practitioner as needed. Even the Homeric epics, originally composed and performed orally, were best learned from and preserved in books.

Socrates can't have it both ways. The idea that disciplines that rely on writing can be praised while the books that enable their practice are denigrated falls flat. And we actually see this idea in the exchange with Phaedrus. As Socrates works to debunk Lysias's speech, he asks Phaedrus to read and quote.

To analyze and critique, the teacher makes recourse to the book. He requests that Phaedrus read from the speech to interact with it. The portable, physical artifact aids the work of criticism just as it serves the purpose of recall. Specific expressions and formulations skirted the mind, but the written word was sitting right there. "Read out from the scroll," says Socrates. It allows not only the first hearing of the speech but also subsequent review and analysis: Take and read.

So the book can be a support to memory, including deeper philosophical consideration, as Socrates's own practice seems to show. What's more, having recourse to quotably precise language aids interaction with the ideas expressed by that language. That acknowledgment doesn't eliminate Socrates's complaint—some of which will prove persistent, as we will see. But it does highlight tensions that drive the drama in the coming pages.

Marginalia: The Genesis of Writing

Most of what we know or assume doesn't come from direct experience. We rely on reports of other people's experiences (or reports of those reports). If a hunter knows about a high-yield game trail or a superior method for felling prey, that information vanishes if he's trampled by his quarry or eaten by something else. It's the same regarding what mushrooms to forage, what to forgo, or any other situation in which a person would rather know than not. Information stored in one person's head, unless shared, dies when they do. So we pass it along.

Sharing ideas is why parents raise children, elders train youths, masters instruct apprentices, priests initiate acolytes, and so on. We call that process of sharing and the body of knowledge it produces *tradition*. It's how our ancestors learned to hunt, fish, and farm and read trail spoor, stars, and seasons. It's also how foundational ideas about being human were first gathered, spread, and preserved. Under these conditions, people learned to knap stone, tan hides, make flutes, build homes, adorn tombs, raise crops, bake bread, rhapsodize, philosophize, and more.

But tradition had (and has) limits, especially if oral transmission is the primary means of communicating from one generation to the next. Following Francis Bacon, we say that knowledge is power. But knowledge is actually rather flimsy. What if the teacher's memory fails or he dies too soon? What if the student is absentminded or dim? If knowledge is general or generally known, the risks are few. One baker is as good as the next, more or less. But the more specialized and narrow the knowledge, the greater the odds and cost of the loss.

Institutions arose to offset that risk. Though susceptible to the usual calamities—and sometimes their cause—institutions are usually more robust than individuals and better at withstanding disruptions. So tribes had elders, kings assembled councils, priests built temples, and teachers started schools, all with a view toward preserving and perpetuating their various traditions—which is to say, bodies of knowledge about the world, how it works, and what we're supposed to do with that ever-growing pile of information. The same motivation drove the development and adoption of writing.

At the dawn of agriculture in the Ancient Near East ten to fifteen thousand years ago, yields were small, and so were communities. People could manage production by word of mouth. But as years passed, surpluses and societies grew in size and complexity. Determining who owed what to whom became crucial—especially as labor specialized, merchants multiplied, and rulers insisted on a cut. The old solutions couldn't keep up, but writing provided an answer.

Cave walls reveal our forebears' mind for art and symbols. But neolithic hunters and shamans don't get the credit for writing. Accountants do. People had long used tally marks in counting. With some imagination, those tallies became inventories, receipts, and records. With a little more imagination, those marks became signs, corresponding to real stuff: cattle, grain, fish, houses, and more. Recognizable writing appeared between five and six thousand years ago as the range of marks expanded to encompass the nuances and complexity of spoken language.

With signs evoking first things and later sounds, ancient scribes could do more than count sheaves and sheep (though that was pretty useful). They could now post declarations, laws, and other inscriptions that asserted royal authority even (or especially) in the king's absence. Writing became a political tool and spread by a succession of empires—Sumerian, Akkadian, Hittite, and others—that regulated peoples and taxed their trade. Writing similarly empowered priests, whose texts enshrined knowledge for appeasing deities, exorcising demons, reading omens, glorifying gods, healing sickness, counseling rulers, and other must-dos. Depending on the context, priests and scribes were synonymous.

A slow but certain revolution followed in writing's wake—slow because writing is tremendously difficult to learn (same with reading). Unlike with speech, there's nothing natural about it. Acquiring the skill took extraordinary effort, especially since the first writing systems were nonalphabetic and necessitated mastering hundreds of signs for fluency. As a result, literacy was first limited to the social ladder's upper rungs, where an underclass of laborers and slaves provided elites with enough leisure for the pursuit. It was also slow because new practices require resounding need and obvious gains to catch hold. For most people, sharing information by word of mouth worked just fine. Most, but not all.

Merchants and traders spotted benefits early on. They found writing helpful for accounting, as did royal and priestly scribes. They also relied on writing for long-distance communication between partners and customers. Shuttling goods to Kanesh (in modern-day Turkey), Assyrian tin and textile merchants used writing to coordinate deals and smuggling runs. Dodging taxes and grabby officials, they gouged cuneiform letters on clay tablets with schemes and logistics, anticipating interference and planning for contingencies.

One ruse found in a four-thousand-year-old letter involved splitting a large load of tin into small packets and having traffickers sneak them into Kanesh "concealed," as one letter put it, "in their underwear." Writing could support or subvert established power. In fact, it was outsiders like these, the Phoenicians, who first spread alphabetic writing, simplifying literacy and expanding its acquisition.

But what made the revolution certain? Writing proved immensely useful, and dependence grew as the benefits emerged. Memory offered the first gains. Writing captured thoughts that might otherwise prove hard to retain or pass along; it was the external hard drive of the mind, cloud storage for your head. This fact is clear from its first uses as record keeping. Who could be expected to remember how many slaves the king possessed or how much grain he dispensed?

Extensive details of any sort evade recollection. There's a reason the poet asks for the Muse's help in the *Iliad* when reciting the lengthy catalog of ships and captains. Even if he had engrained the list in his brain through practice, the person listening could never repeat the list with accuracy after one or two hearings. So ancient libraries filled with all sorts of reference works for consultation by those who needed information they couldn't easily recall on their own.

But the benefits of writing went—and go—beyond mere memory. Writing leads to editing, which is to say writing leads to thinking. More on that topic next.

Chapter 2

UPGRADING THE MIND

The Magic Stroke

In 1981, archaeologists unearthed a tomb just south of the Acropolis in Athens. The grave dated to between 430 and 410 BCE. It contained a skeleton, iron tools, musical instruments, and writing implements: a small case with bronze ink pot and bronze stylus, five small writing tablets, and a roll of papyrus. While the identity of the past owner remains a mystery, scholars identify the person as a poet, a musician, possibly a teacher.

The grave items have been subject to study in the decades since their discovery. Three of the writing tablets were tied together along one vertical edge to form a notebook. The tablets and papyrus seemed to have contained epic or lyric verse. A few letters and words are still legible and represent the oldest known examples of Greek writing on tablets or papyrus.

The writing reveals a skilled—though delightfully imperfect—practitioner. On one line, as an analysis of the tablets says, "there appears to be an erasure" with a correction "written above in the interlinear space." Equipped with devices featuring delete keys and autocorrect, we might consider this finding unremarkable. But it's a tantalizing indication of the underlying power of writing.

Writing tablets were employed in the ancient world and through the medieval period. The format was near universal: a recessed wooden (sometimes ivory) tablet, forming a shallow tray. Filling the recess with wax created a malleable surface on which letters could be incised with a stylus. A writer could inscribe the words she wanted, smudge out those she didn't, and write over the erasure—the ancient precursor to the word processor.

Among the literate, these waxen tablets were almost as ubiquitous as smartphones are today. Power users bound three, four, or more tablets together, multiplying writing surfaces and expanding their creative capacities. Writers like the Attic Poet favored them because they facilitated easy composition, including editing. And to facilitate that is to facilitate critical, analytical, constructive thought—something Socrates tacitly acknowledged in his own behaviors, if not his words.

Consider Socrates's two-column test from before. Having written his answers across columns, Euthydemus's faulty judgment now stares

First-century portrait of a woman holding a set of wax tablets and stylus. *SOURCE*: MUSEO ARCHEOLOGICO NAZIONALE DI NAPOLI. AZOOR PHOTO, ALAMY.

him in the face. Socrates could have led him through the same process without writing. But by rendering the lesson visible, he drew on more of Euthydemus's cognitive and critical faculties.

Writing externalizes ideas, rendering the ideation process itself observable. Subjective thoughts become objects that can be analyzed, challenged, dissected, rearranged, and revised. The Attic Poet's correction in wax represents a record of her thinking, seen after the fact. But it is more than a record of mental effort. The articulation and visual display of the original thought encouraged the challenge and insertion of the second. Beyond merely representing thinking, in other words, writing *is* thinking.

"The magic of writing arises not so much from the fact that writing serves as a new mnemonic device," says David Olson in *World on Paper*. Helpful as that might be, there's more going on. "Writing," he says, "not only helps us remember what was thought and said but also invites us to see what was thought and said in a new way." As philosopher Andy Clark argues, using the tools of writing—pen, paper, keyboard, screen, whatever—forms a circuit outside the body that extends the functionality of the mind. Without that extended circuit, our thinking is diminished and produces less in the way of ideas.

Writing thus sparks and supports heightened cognitive engagement, and we can sometimes glimpse how impactful this was to the writers themselves, how it shaped their working habits along with the work they produced. Let's start with the Roman poet Vergil.

How the Pros Do It

When Vergil wrote his poem *Georgics*, he would dictate lines all morning. His assistant would copy them down, and the two would zhuzh and tweak, rearrange and delete all afternoon, until what remained was satisfactory. "He fashioned his poem after the manner of a she-bear," said biographer Suetonius, "and gradually licked it into shape."

A writer or editor might choose a different image today, but they perform the same function in their work. The process allows expression to move from approximate to exact, inadequate to ample, partial to complete, clunky to eloquent—all of which depends on the epistemological gains from thinking with the assistance of an external medium, such as a wax tablet, parchment notebook, or keyboard and screen.

We can assume Vergil followed a similar process when he wrote the *Aeneid*, the famed poem of Rome's founding, meant to serve as a national epic on par with the *Iliad*. "After writing a first draft in prose and dividing it into twelve books," says Suetonius, again taking us behind the curtain, "he proceeded to turn into verse one part after another, taking them up just as he fancied, in no particular order. And that he might not check the flow of his thought, he left some things unfinished, and, so to speak, bolstered others up with very slight words, which, as he jocosely used to say, were put in like props, to support the structure until the solid columns should arrive."

Vergil works from multiple drafts to ensure that he covers what he wants and has the core ideas down before he tries to polish them. He uses sketches and notes to mark portions for later attention. He gets cursory thought down so he doesn't forget it amid the flow of composition, knowing he'll need to return and flesh out his ideas later. The jots and scratches dropped at later points in the narrative work like mental scaffolding, allowing him to hold on to thoughts that might otherwise escape him. And his compositional medium—tablets or parchment notebooks—permit him or his assistant to jump back and forth as needed throughout the story.

While telling the story of unlucky Byblis drafting a love letter, the Roman poet Ovid provides another picture of this reflective process in action. "She proceeds to set down with a trembling hand the words she has thought out," he says. "She begins, then hesitates and stops; writes on and hates what she has written; writes and erases; changes, condemns, approves; by turns she lays her tablets down and takes them up again."

These external gestures telegraph internal movements. The back-and-forth with her tablet both reveals and aids the dance in her mind. Byblis thinks and rethinks and uses her externalized thought to improve the thought and its expression. This word, not that. Byblis's effort to articulate ideas on wax becomes an inseparable part of her ideational process; her

The unlucky Byblis, writing a love letter to her brother, who is understandably put off. Engraving by Crispijn van de Passe. SOURCE: RIJKSMUSEUM. WIKIMEDIA COMMONS.

ideas never take their final shape without the aid of writing because she requires the critical interplay afforded by visible words.

How could Ovid imagine a writer venturing for clarity by writing and rewriting until just the right phrase emerged from the fog of possible choices? Undoubtedly because that's how Ovid approached his own work. And, of course, we have surviving physical evidence indicating that's how the Attic Poet worked.

Writing catches ideas in flight. Inscription fixes them long enough to inspect. Then we're free to scrap them, settle on them, or even sanctify them. And when we return to them at a later point, we find they are still fixed conveniently where we left them. That was the key to Ovid, Vergil, and the Attic Poet. They wrote what they thought and then rethought what they wrote.

The Technological Side

This critical engagement with solidified sounds, shaping words like a sculptor—to use an image from Roman orator Quintilian—facilitated more complete and complex thought. Quintilian draws this idea out more clearly than most. "We must first criticise the fruits of our imagination, and then, once approved, arrange them with care," he said, going on to say,

> We must select both thoughts and words and weigh them one by one. This done, we must consider the order in which they should be placed, and must examine all the possible varieties of rhythm, refusing necessarily to place each word in the order in which it occurs to us. In order to do this with the utmost care, we must frequently revise what we have just written. . . . We love all the offspring of our thought at the moment of their birth; were that not so, we should never commit them to writing. But we must give them a critical revision, and go carefully over any passage where we have reason to regard our fluency with suspicion.

"Murder your darlings," Arthur Quiller-Couch quipped in 1916. Quintilian said the same nineteen centuries prior.

It's possible to compose in the mind without writing. The most ancient compositions were oral, usually poetry; those we still have today were later captured in writing, sometimes flattened into prose. But composing by way of writing does make the process easier and extends the capability. Even accounting for variability in working memory and long-term memory for those who have trained them both, writing offers advantages.

As it happens, Quintilian recommends wax tablets for this work. A writer might use parchment as well, he says, but moving to and from the inkwell slows a person down; the stylus in wax eliminates that step. (This idea complicates the trendy, though minority, opinion that writing by hand trumps digital word processors. Supposedly, slower writing produces more thoughtful writing. Yet here's Quintilian, one of the premier talents of his day, suggesting the opposite.)

Modern archaeology does provide evidence of composition and editorial work beyond wax tablets alone. Using advanced imaging technology, scholars have deciphered parts of a papyrus scroll from a library in Herculaneum, carbonized two millennia ago in the eruption of Mount Vesuvius. A working draft of philosopher Philodemus's history of Plato's Academy, the scroll contains not only the primary text but also the author's marginal ideas for inclusion in the final draft. "We can see he made notes—insert this later, insert this and this, and skip this and this for the final version," says University of Würzburg professor Kilian Fleischer of the discovery.

Whatever the medium, Quintilian stresses formatting the writing space with enough room for corrections and additions. The tool affects the product. Quintilian recognizes that writers have random thoughts that require immediate capture for later use. "Sometimes," he says, "the most admirable thoughts break in upon us which cannot be inserted in what we are writing, but which, on the other hand, it is unsafe to put by, since they are at times forgotten, and at times cling to the memory so persistently as to divert us from some other line of thought. They are, therefore, best kept in store."

This is very similar to Vergil's approach, using the writing medium to keep track of inchoate ideas for subsequent development. In purely oral composition, a poet, historian, or orator would have to hold those in-breaking thoughts somewhere in their working memory while they settled the problem at hand. This is tricky because working memory is notoriously short. Writing allowed people to outsource that function to a tablet or page while they reserved their focus for their primary task.

Quintilian shows us the technological side of writing and what it means for our ability to shape our thoughts and share them with others. Through the lens of Quintilian, Vergil, and others, we see that the tools of composition impact not only the thinking process but also the thoughts composed.

M FABIVS QVINTILIANV VICTORIO MARCELLO S D SAL

FFLAGITASTI QVOTIDIANO CON
VICIO VT LIBROS QVOS AD MARCELLV
meum de institutione oratoria scripserim iam emittere
inciperem. Nam ipse eos nondum opinabar satis matur
uisse. Quibus componendis ut scis paulo plus q̄ biennii
tot alioqui negotijs districtus impendi quod tempus non
tam stilo q̄ inquisitioni instituti operis prope infiniti & legendis auto
ribus qui sunt innumerabiles datum est usui. Deinde Oratij consilio q̄
tirce poetica suadet ne precipitetur editio nonum̄q̄ prematur in annum
dabam ijs otium ut refrigerato inuentionis amore diligentius repetitos
tanq̄ lector perpenderem. Sed si tantopere effiagitantur q̄ tu affirmas p̄
mittamus uela uentis & oram soluentibus benepreceremur; Multum a͞t
tua q̄ fide ac diligentia positum est ut in manus hominum q̄ emendatis
simi ueniant;

M F Quintiliani institutionis oratorie ad Victoriū Marcellū pmus liber incipit cap̄
I Quenadmodum prima elementa tradenda sint. II Verum utilius domi
an scolis erudiantur. III Qua tōne in pueris ingenia dignoscantur et
qui tradenda sint. IIII De grāmatica. V De officio grāmatica. VI An o
ratori futuro necessaria sit plurium artiu sciētia. VII De musica. VIII
De geometria. IX De prima pronuntiationis & gestus institutione
X An plura eodem tempore doceri prima etas possit;

M F Q POST PROHEMIVM IN LIBRIS DE INSTITVTIOE
POST imperatam studijs meis quietem quam plurimi annos e ORATORI
erudiendis iuuenib; impenderam cum a me quidam familiariter
postularent ut aliquid de ratione dicendi componerem diu sum eqdem
reluctatus q̄ auctores utriusq̄ linguę clarissimos no ignorabam multa q̄
ad hoc opus pertinent diligentissime scripta posteris reliquisse. Sed qua e
go excausa faciliorem mihi ueniam meę deprecationis arbitrabar fore
hac accendebantur illi magis q̄ inter diuersas opiniones priorum & quasdam
etiam inter se contrarias difficilis est electio ut si mihi no inueniendi noua
At certe iudicandi de ueteribus ingerere laborem non iniuste uiderentur
Quamuis a͞t non tam me uincerer prestandi q̄ qd exigebatur fiducia q̄
negandi uerecundia. Latius se tamen aperiente materia plus q̄ iponebat
oneris sponte suscepi simul ut pleniore obsequio de me mererer ama
tissimos meq̄ simul ne uulgarem uiam ingressus alienis demum uestigijs
insisterem. Nam cetteri fere qui artem orandi litteris tradiderunt ita

Sixteenth-century manuscript of Quintilian's *Institutio Oratoria*. SOURCE: BIBLIOTECA APOSTOLICA VATICANA, VATICANUS PALATINUS LAT. 1556, FOL. 1R. WIKIMEDIA COMMONS.

Using the Extended Mind

It's difficult to imagine this sort of composition in a purely mental space or oral setting. Artifacts of oral composition reveal differing priorities and abilities of their composers than those held by the composers of original written texts. In composing the *Iliad* and *Odyssey*, for instance, Homer and his bardic heirs prioritized memorability over detail and variety. So we find the use of repeated formulations and mnemonics—"swift-footed Achilles" and "wine-dark sea"—a technique Vergil's written composition has no need for. Not only is the memory of the final product outsourced to the page, but so is the mental effort of the composition itself, allowing a level of complexity practically impossible for the memory alone to maintain.

The ancients memorized massive amounts of information, but contemporary studies show oral compositions are inexact and subject to alteration, both intentional and accidental. This means that a given oral text not only has no final, canonical, or official version but also is subject to potential revision on every recitation. It likewise means poets were constrained by their medium—the mind itself—to eschew the sort of extensive detail and complicated exposition that resists unaided recall.

The act of writing changed the quality and scope of ideas. Externalizing thoughts on a visible medium such as wax or parchment augmented the thought process itself. The tool became part of the thinking. "Something ineffable happens when you write down a thought," says historian Lynn Hunt, adding, "The process of writing itself leads to previously unthought thoughts. Or to be more precise, writing crystallizes previously half-formulated or unformulated thoughts, gives them form, and extends chains of thoughts in new directions." That is, it moves thought up the *y*-axis.

Hunt speculates that writing triggers an interaction between our mind and the medium of composition—"a kind of shifting series of triangulations between fingers, blank pages or screens, letters and words, eyes, synapses or other 'neural instantiations.' . . . By writing, in other words, you are literally firing up your brain and therefore stirring up your conscious thoughts and something new emerges."

Neuroscience bears this idea out. Functional magnetic resonance imaging reveals noticeable engagement in brain areas known for creativity in comparisons of people reading, thinking, *and writing* with those who are merely reading and thinking. Writing in general promotes heightened creativity, and practiced writers show amplified results above and beyond nonwriters.

Writing allows you to review what you have thought, to retain what you have thought long enough to review it. This process facilitates iterative, constructive thinking. One thought can be added to another—and another and another. Through the magic of editing and revising, writing allows you to insert new thoughts between the old ones or to erase fruitless lines of argument or narrative.

Building complex arguments or stories requires the ability to retain large amounts of information in your mind unless you can delegate the retention to a document, which then makes it available for editing, interpolation, refinement. Writing thus encourages the generation of increasingly complex ideas. Arguments that require separate lines of inquiry, contingent details, proofs, and examples can be layered and rearranged to strengthen the claims. Moreover, it lets the poet change the colors of her palette, even those already on the canvas. As a result, writing permits a level of complexity, precision, and eloquence in communication that is unattainable by word-of-mouth efforts.

Precise formulation enables the thinker to go beyond simple ideas and tackle intricate theories. It allows the mind to solve one problem and, having retained it in writing, turn to another. Individual elements of interrelated problems can be tackled one at a time and then addressed as a whole. Vergil using stub verse as mental props and Quintilian saving space for random thoughts demonstrate solutions to compositional challenges of putting thoughts in order.

Plato's Secret

Owing to cognitive limitations of memory, linearity of thought, and other challenges, Vergil, Ovid, and Quintilian could never have formulated their ideas without writing. Nor could Plato have rendered his many portraits of Socrates without resorting to the technique and technology about which his mentor appeared so ambivalent.

Plato built his arguments one word at a time just like the rest of the writers we've encountered, and he was also known to polish and refine after the fact, using—what else?—a wax tablet. In a treatise on rhetorical methods, written around the turn of first millennium, Dionysius of Halicarnassus tells us, "Plato did not cease, when eighty years old, to comb and curl his dialogues and reshape them in every way. Surely every scholar is acquainted with the stories of Plato's passion for taking pains, especially that of the tablet which they say was found after his death, with the beginning of the *Republic* ('I went down yesterday to the Piraeus together with Glaucon the son of Ariston') arranged in elaborately varying orders." The philosopher whose mentor supposedly denigrated writing left evidence of his use of this very technique, including zhuzhing and tweaking to get his expression just right.

Knowingly or not, Plato availed himself of all the cognitive gains writing made possible. So do all writers. As we've now seen, writing is more than a record of thinking. It is a form of thinking unto itself and one necessary for elaborating detailed formulations and encoding them with the kind of precision necessary for readers to engage those formulations and construct new ones of their own.

Marginalia: Notes to Self

Sidonius Apollinaris, the aristocrat and bishop we met earlier, was generous to a fault and a little irked by the result. "The most learned man in the world" in Sidonius's estimation, one Mamertus Claudianus, had written a dazzling new book, *On the Nature of the Soul*. Sidonius owned a copy but loaned it to his friend Nymphidius before he finished using it. Bad move. Then as now, no one ever returned a borrowed book. And, sure enough, Nymphidius was still sitting on it long after Sidonius expected its return.

"It is high time for you to send the book back," he wrote. "If you liked it, you must have had enough of it by now; if you dislike it, more than enough. Whichever it be, you have now to clear your reputation." Sometimes it takes a little shame to motivate people. But why did Sidonius want the book back anyway? Indeed, what was Nymphidius doing with it?

The human mind is capacious and capable of remarkable feats of memorization, especially when utilizing ancient mnemonic techniques, such as, say, building a memory palace—that is, arranging information in the mind as if in a real space that an individual can visit in their mind's eye and, by going from one location to another in the imaginary space, call up the desired memories associated with each installation or feature. But, as we've seen, memory is faulty and no match for extensive detail or the intricacies of argumentation—exactly the kind of thing Sidonius and Nymphidius encountered in Claudianus's book.

"In this book," said Sidonius, "Grammar divides and Rhetoric declaims; Arithmetic reckons, Geometry metes; Music balances, Logic disputes; Astrology predicts, Architecture constructs; Poetry attunes her measures." Nobody's keeping all that straight, even if they try.

No problem for Sidonius, of course. He could always check his copy and look up any passage he desired for a refresher—if he could ever get it back from Nymphidius. But then what would Nymphidius do, especially once he'd returned the book? The letter answers that question as well, along with explaining why Nymphidius borrowed the book in the first place—not merely to read it, as it turns out.

"Pleased with the novelty of a theory like [Claudianus's], and kindled to enthusiasm by so much ripe wisdom, you had hardly seen the book before you asked to have it for a short time," said Sidonius. Why? "To examine and copy it and to make extracts." By making extracts, Sidonius meant copying choice bits and pieces to retain after he'd returned the book. This served the same function as having the book itself, except perhaps allowing more direct access to the desired thought (though sacrificing the immediate context).

Modern researchers sometimes call working memory the sketchpad of the mind, but working memory goes only so far. Extracting stretches the brain's sketchpad, helps someone integrate the ideas into long-term memory, and, importantly, helps them find those ideas again if long-term memory disappoints, which it will. By borrowing Claudianus's book and harvesting its choicest lines and passages, Nymphidius was extending the capacities of his own memory, directly and indirectly. In his study of commonplace books (notebooks designed and arranged to capture such memorable ideas), Earle Havens refers to this process as "artificial memory."

Beyond that, however, Nymphidius was expanding his cognitive repertoire for generating new thoughts. The Latin grammarian Macrobius, an exact contemporary of our two friends, described both benefits:

> We ought in some sort to imitate bees; and just as they, in their wandering to and fro, sip the flowers, then arrange their spoil and distribute it among the honeycombs, and transform the various juices to a single flavor by some mixing with them a property of their own being, so I too shall put into writing all that I have acquired in the varied course of my reading. . . . For not only does arrangement help the memory, but the actual process of arrangement, accompanied by a kind of mental fermentation which serves to season the whole, blends the diverse extracts to make a single flavor; with the result that, even if the sources are evident, what we get in the end is still something clearly different from those known sources.

As we'll see later, Macrobius stole this idea from Seneca—notably the image of the bee. But he also developed it further. Jotting down the choice words of others both heightens recall and assists in generating novel thoughts by combination, refutation, or inexplicable leaps. Macrobius's version of this insight is more robust than Seneca's. New connections, new relationships, new challenges emerge as we contemplate the ideas of others, especially if we're doing so deeply and thoroughly enough to be taking notes on what they're saying, extending the native capacities of our unassisted minds—and jotting down not just the gist of the ideas but actual quotable lines and passages, something that allows us a level of intellectual precision otherwise unavailable to us.

This insight applies beyond writers and includes any reader willing to pause long enough to record their thoughts, either inside the book as marginalia or outside on another surface, such as (then) wax tablets, parchment or papyrus scrolls, codices, scraps, or (now) paper notebooks, mobile apps, laptops, tablets, and e-ink devices, or whatever else comes easily to hand. That is to say, interacting with the written word by writing words enhances our ability to both remember and generate new ideas. And, as the next chapter shows, we've only just scratched the surface in our story so far.

Chapter 3

TOOLS FOR THINKING

From the Archives

After King Tut died in 1325 BCE, the Egyptian throne was empty. His young widow, Ankhesenamun, needed a new husband in a hurry. Royal officials were scheming. They wanted to marry the queen, marginalize her, and seize the title of pharaoh for themselves. So Ankhesenamun dispatched a messenger north to Suppiluliuma, king of the Hittites.

"My husband is dead, and I have not a son. But," the resourceful widow said in a letter, "they say that you have many sons. If you should give me a son of yours, then he shall be my husband."

Suppiluliuma was suspicious. The rival kingdoms, Egypt and Hatti, had warred for years. Surely this message was a prank. "No doubt they are making sport of me," he said before sending an emissary of his own to bring back "reliable information." Irked by the suspicion and delay, Ankhesenamun wrote again, now more insistently than before. Time was short, and her fate would soon be sealed if the Hittite king denied her a royal partner. But Suppiluliuma was still undecided. What should he do?

A biography of the king, written by his son, says what happened next: "My father asked for the treaty tablet [which described] how formerly the Stormgod had . . . established a binding bond between Egypt and Hatti . . . how they were forever friends among each other." The aged marks incised in clay were clear. Should Suppiluliuma give Ankhesenamun one of his sons? Should he ally his kingdom with Egypt? The tablet provided grounds for doing so.

What was true in the past would be true again! Suppiluliuma agreed to send his boy. "In the old days Hattusa and Egypt were friends among each other," he said, "but now this too has happened between us, so that Hatti and Egypt forever after will be friends to each other!" Alas, no. The marriage never occurred. The Hittite prince was murdered en route, Ankhesenamun was forced to marry one of the schemers, and the kingdoms went to war.

The story of the widowed queen and murdered prince contains all the elements of grand narrative: palace intrigue, betrayal, homicide. But one scholar of ancient languages, Theo van den Hout, noticed something else—namely, that clay tablet, written almost a century before, filed away, and called up when circumstances required.

Egyptian-Hittite treaty in cuneiform. This example dates to about seventy-five years after the one consulted by Suppiluliuma.
SOURCE: NEUES MUSEUM, BERLIN, GERMANY. OSAMA SHUKIR MUHAMMED AMIN, WIKIMEDIA COMMONS.

"From this and many other passages it becomes clear that Hittite kings were able to order specific tablets from their tablet rooms," he says. "Either the original or a copy of the original must have been filed somewhere and was retrieved from the tablet rooms." That might seem unremarkable in the age of Google and large language models, when knowledge of any kind awaits only a few keystrokes and a click. But in the second millennium BCE, such recall was astounding.

Search Engines

Ancient archives and libraries formed an analog internet for their users, and the clay tablets and waxen writing boards they housed were among the world's first information technologies. Properly organized, libraries could store hundreds, even thousands of documents for use; Suppiluliuma's trove in Hattusa contained as many as seven thousand, categorized, searchable, and available as needed.

Filing and retrieving information was essential for administering ancient kingdoms and empires. The earliest and largest libraries were housed in palaces and temples and featured the sorts of books royal and religious officials might need for reference: records and annals, omens and oracles, legislation and lexicons, prayers and myths, spells and incantations.

Officials employed them as we might today: to pass along important information, confirm elusive facts, and answer thorny questions. Rulers saw library building as an important task. Thus the Seleucid King Ptolemy I founded the library of Alexandria, the Attalid King Eumenes II founded the library of Pergamum, and the Judaean Governor Nehemiah founded a library in Jerusalem, which Judah Maccabee rebuilt after its destruction.

To build their libraries, rulers resorted to patronage, appropriation, even conquest. Few ancient monarchs pulled off this mix as memorably as Ashurbanipal, who ruled the Neo-Assyrian Empire about twenty-seven hundred years ago. A ruthless lord who humiliated his enemies and killed lions for sport, Ashurbanipal also boasted of his own literacy and hoarded books.

When archaeologists unearthed Ashurbanipal's library in the nineteenth century, they found omen collections, rituals, hymns, histories, medical treatises, language instruction, myths, and more. In all, they discovered between fifteen hundred and several thousand works on thirty thousand clay tablets—by far the biggest library of its time. Acknowledging the incongruity between Ashurbanipal's violent rule and his literary bug, one museum curator called him a "psychopathic bookworm."

While Ashurbanipal acquired books through peaceful means, such as inheritance and hiring scribes, he also seized holdings as war booty, confiscated them from subjects and foes alike, and enslaved foreign scribes to work in chains. In a rare feat for a king, he also copied books himself. "I wrote on tablets," he said in one inscription, "and deposited [them] within my palace for perusing and reading."

What drove his acquisitiveness? A letter Ashurbanipal sent to underlings working in Babylon illuminates at least one motivation. "The day you see my letter," he said, "collect all the tablets [in the palace] and all the tablets . . . stored in the temple." He then listed several specific books of omens, rituals, and instructions he desired. "Search for them and bring them to me!" he said, before expanding the order: "As for any tablet or instruction that I did not write to you about but that you have discovered to be good for the palace, you must take [them] as well and send [them] to me."

Rulers like Ashurbanipal equated data with power. Understanding omens meant greater confidence and better decision making. Correctly following the proper ritual for this or that purpose meant more reliable outcomes, or so they believed. Thus, the more books on such topics he possessed, the more secure his rule.

The primary purpose of such royal libraries, says Ancient Near East scholar Eleanor Robson, "was to provide large datasets of omens to aid royal decision-making and rituals to ensure continued divine support for the crown." That seems so by the numbers. Based on itemized lists of tablet contents, royal libraries were preoccupied with omens and divination.

Ashurbanipal on a lion hunt. Alabaster bas-relief. *SOURCE:* PERGAMON MUSEUM, BERLIN, GERMANY. GARY TODD, WIKIMEDIA COMMONS.

But in whatever library built for whatever purpose, the holdings dealt with descriptions, prescriptions, and proscriptions whose details were too elusive, extensive, and varied to entrust to memory alone. Whether preserving a simple list, complicated procedure, intricate recipe, or captivating stories, the book harnessed thoughts—all so people could think about them later (the z-axis).

Because of this preservation, old ideas and new perspectives could be brought to bear on future considerations. Ancient myths and novel tales alike could explain the world, even the human heart, to itself from one era to another. And all these externalized fragments of memory could be herded, gathered, and amassed in collections that powered more thinking, more imagination, more ideas—conceptions beyond those originally conceived but dependent on the original as a scaffold or spark for something new.

Economist Tyler Cowen illuminates the basic idea. "Once an idea has been generated," he says, "it can be used many times by many different people at very low marginal cost." Instead of generating knowledge from scratch—which is costly, sometimes even deadly—we can work with ideas already in existence. And one of the best means for accessing those ideas is a book. As literary critic I. A. Richards said, "A book is a machine to think with." The only thing better? More machines.

Plato's R&D Department

Despite his teacher Socrates's ambivalence about books, Plato was a surprisingly eager user. He not only wrote books but also engaged with the ideas of others through their books and built a library of his own. In his *Lives of Eminent Philosophers*, for instance, Diogenes Laertius reports that Plato traveled to what is now Italy to visit certain Pythagorean philosophers. His interest in their ideas persisted, and he began to acquire Pythagorean books at significant cost. "Some authorities, amongst them Satyrus, say that he wrote to [his disciple] Dion in Sicily instructing him to purchase three Pythagorean books from Philolaus for 100 minae," says Diogenes. We needn't fret about the expense. "For," as Diogenes explains, "they say he was well off."

On at least one occasion Plato accepted payment in books instead of money. But, of course, money helped too; books weren't cheap, and Plato wanted many more. Why? They formed a vital part of his intellectual life, not only for personal edification but also for theorizing, disputation, and critique. The books of earlier writers and contemporaries provided the raw material for his own thinking and development. Beyond the Pythagoreans, for instance, Plato borrowed ideas from the books of poets Epicharmus and Sophron; he supposedly tucked a copy of the latter under his pillow at night.

Whether by adopting and adapting ideas or dissecting and rejecting them, Plato improved his own thinking by accessing that of others through books. In one comprehensive study of available biographical data and Plato's own writings, architect and book historian Konstantinos Staikos counted nearly sixty works Plato either owned or likely owned, along with almost forty more he cited or whose ideas he commented on. That's nearly a hundred titles one could expect to find housed in the library of his Academy outside Athens, not to mention a couple dozen more copies of his own writings.

And the number was probably much higher. The likely ruins of the Academy were unearthed in the middle years of the twentieth century. Based on a reconstruction by Staikos, we can imagine a high degree of literary industry: student noses slowly plowing through scrolls; a reader performing texts aloud, while eager ears crane close and skeptical eyes dart around, gauging the book's reception by the looks of peers and mentors; finally, students sitting at stations for copying manuscripts, destined to fill shelves in the Academy library, a cubby in the student's lodgings, or the study of one of their patrons. Simon Critchley, a philosophy professor who specializes in classics, describes Plato's library as his "research engine."

No record exists of these books or what happened to them when Plato died, but he likely bequeathed them to his successor at the school. If we imagine that successors and students added books of their own, the scope becomes huge. Within just a generation of Plato's death, Staikos says, the Academy library would have been the biggest around—though not for long.

Aristotle's Method

Plato's equally famous student Aristotle differed from his teacher in many ways, but he also loved books. Plato, who nicknamed Aristotle "the Mind," appointed him "reader" in the Academy after his arrival in 356 BCE. Prior to the modern world of print and mass distribution, reading a book aloud before an audience was a key step in publishing its contents. It was also part of the philosophical profession—listening to someone's arguments read aloud, picking praiseworthy points and critiquing unsuccessful lines of argument; it's exactly what we see Socrates doing in Xenophon's telling, and it's similar to what we see in Plato's *Phaedrus*. As we'll observe later, reading was thought arduous and often reserved for slaves. Aristotle, who was a person of means like his teacher, nonetheless enjoyed the chore and passed the joy to his audience. When Aristotle was absent from school, Plato remarked, "The lecture room is dull." If Aristotle possessed a love for books before joining the Academy, his time there cemented the passion.

"The Mind" caught other nicknames from his teacher. Plato compared him to a colt, because he bucked his mentor's expectations, ultimately leaving and founding a rival school, known as the Lyceum, where his bookishness rose to new heights. Aristotle's library at the Lyceum has also been discovered. Construction workers located the site in 1997 while digging up a parking lot roughly a quarter mile from Athens' Constitution Square. They were preparing the ground for a new museum of modern art; instead, they found a missing link to modernity itself.

Aristotle's writings attest to the centrality of books in his method. In his *Topics*, for instance, he advises researching prior authorities and opinions before formulating ideas of one's own: "We should select . . . from the written handbooks of argument, and should draw up sketch-lists of them [for each individual subject], putting them down under separate headings, e.g. 'On Good,' or 'On Life.' . . . One should indicate also the opinions of individual thinkers, e.g. 'Empedocles said that the elements of bodies were four': for anyone might assent to the saying of some generally accepted authority."

Where to begin an inquiry? For Aristotle, the answer lay in reviewing the available literature. Existing books helped the inquirer shape their understanding of a subject, especially as they began to work with the ideas by making and manipulating lists and notes. The library provided paths for exploration.

Working with books was valuable enough that the lack of written material from which to work proved an impediment. Aristotle said as much in his book *On Sophistical Refutations*. Earlier work on rhetoric was easier, he said, because "there exists much that has been said long ago" (in books, of course). Not so for the subject of reasoning. As a result, "we had nothing else of an earlier date to speak of at all, but were kept at work for a long time in experimental researches." The same was true for his natural science work. Aristotle relied on other authorities when he could and conducted investigations of his own when he couldn't, dissecting cuttlefish and other animals to understand their anatomy. Aristotle recorded his findings to ensure that others could access and advance his own discoveries.

Aristotle demonstrated the value of books in another way as well—by their absence. After he departed Athens, he gave his books to his successor, Theophrastus. Later, when he died, Theophrastus willed the books to a man named Neleus. But Neleus fell out with his peers, left in a huff, and hauled off all the books, including many that existed only in single copies. This unforeseen eventuality bankrupted the school intellectually. Without their library, students "were therefore able to philosophize about nothing in a practical way," said Strabo in his *Geography*. Instead, they "talk[ed] bombast about commonplace propositions."

Ancient Big Data

These lives and stories highlight the value of books in spurring and shaping thought, including the decline of thought in their absence. Given their value and influence, books—like the information they contained—were bound to multiply. And, practically speaking, there's no such thing as just one. Even one-size-fits-all texts like the Bible reveal many sources in their composition and yield many more after their completion. A solitary book is already a library unto itself and others waiting to happen. But multiplying books need management to remain useful.

Relatively small libraries like Plato's could be organized and managed without much trouble. Larger libraries like Aristotle's required more care and intentionality. Any large library needs an organizational plan and regular maintenance to remain useful. What use is a book if it can't be found when needed? We can go back to clay-tablet libraries like those we've already observed to learn how the challenge was initially answered. Ancient archivists dealt with the difficulty in a number of ways. First, the clay tablets themselves: They couldn't be bound like pages in a paper book (that format was still a long way off). Instead, works that took more space than a single tablet would be incised on multiple tablets, labeled, and numbered so they could be kept and consulted in order. A ruler like Suppiluliuma or Ashurbanipal could ask his servants to fetch a particular series of tablets and know he was getting what he asked for. Likewise, if a tablet fell out of sequence or was shelved incorrectly, the label allowed archivists to bring the prodigal home.

Next, tablets were grouped and shelved together based on similar uses and subject matter. Sometimes they were shelved face out for quick and easy consultation. They might be bundled in baskets, stacked on shelves, or tucked into niches and cubbies in the wall: this subject here, that genre there. Libraries of extraordinary size might be subdivided among multiple buildings, as was the case of Ashurbanipal's collection. The physical space of the library had organizational qualities that could be harnessed to arrange and order books.

In small libraries, mere familiarity with the collection was enough to locate, file, and retrieve particular books. But larger collections required more sophisticated information management methods. Libraries kept inventories and even developed full-blown catalogs. The Hattusa catalog contained useful metadata, such as titles taken from the first words of the book and brief descriptions of the contents. Someone could scan the catalog, spot what he was after, and retrieve a single tablet or series of tablets for consultation.

Libraries were meant for use, and acquisitions and turnover presented a challenge. To manage the stampede of Ashurbanipal's acquisitions, his scribes kept incoming titles listed on waxen writing tablets (the very same technology used by the Attic Poet and Vergil long after). As we've seen, they could be used for permanent storage and consultation or erased and overwritten like today's whiteboards or notetaking apps.

Organizing and ordering books helped organize and order the world. Ashurbanipal's library was the Google of its day, an internet of clay and sandal leather, its size, scope, and prestige only eclipsed three and a half centuries later by the Great Library of Alexandria, where the same basic challenge persisted.

How does one organize thousands upon thousands of scrolls? Alexandrian librarians employed different forms of metadata: attaching title tags to scrolls, storing similar books together, and even assembling a massive 120-volume catalog that provided additional data, including the author's name, his father's name, his profession, the number of lines in the work, and so on. What's more, this was all arranged in elaborate tables, alphabetically by the author's name—a novelty at the time and an organizational breakthrough. Historian Dennis Duncan referred to the whole operation as "Greek Big Data."

And it was all for the purpose of making ideas accessible and thus usable. What once lived in the mind of one person could then be used to provoke the thoughts of another, and the combination could add still more knowledge to the project of human understanding. Of course, that assumes a person understands what they're reading. And to understand the challenge there, we will circle around to the eastern side of the Mediterranean. First, however, I want to explore one additional aspect of organizing information and what emerges as we do.

Marginalia: Cicero's Analog AI

Cicero found himself temporarily exiled from Rome in 58 BCE. Within two years, he was back—with a mess on his hands. His property had been ransacked and vandalized in his absence. The damage to his library was particularly grievous. Scrolls were jumbled, scattered, torn, missing their title slips. A library assumes an order, a schema, a program, something that renders it sensible and accessible. Cicero's was chaos. Fortunately, in stepped Tyrannio, a Greek specialist in literature and libraries.

Tyrannio's reputation preceded him. He supposedly possessed a massive library of his own, some thirty thousand scrolls. Additionally, as a known lover of the best in Greek literature (especially Aristotle), he had already played a key role in the transmission of the philosopher's books to the West.

After Theophrastus willed away Aristotle's books, they eventually ended up with people who had no business keeping such treasures. In an example of spectacularly bad timing, the Ptolemies of Alexandria and the Attalids of Pergamum had begun competing over which kingdom could build the more prestigious library, and the undeserving possessors of Aristotle's books got antsy lest their Attalid neighbors discover their prize and seize the books for themselves. So they stashed the books in an underground vault, where they were predictably damaged by water and worm. Damaged but not destroyed.

The books were still valuable, and when the coast was clear, descendants unearthed the lot and sold it to Apellicon of Teos, a wealthy book collector who brought the books back to Athens. Apellicon repaired the damage, though apparently not well, introducing errors and corrupting the books even further. Still, the scholars at Aristotle's school were glad to make copies and restore their collection. Now they were, as Strabo says, "better able to philosophize and Aristotelize," though with reservations because of the corruptions.

They had other things to worry about as well. By then Rome was up and knocking over its neighbors. The Roman general Sulla invaded Athens and seized Apellicon's library, including all the books of Aristotle. He dragged the collection back to Rome as war booty. That's where Tyrannio got his hands on them after schmoozing the official librarian placed in charge of the books. The books were practically unknown, but, according to Plutarch, Tyrannio worked to restore the collection and eventually ensure their wider publication. "It is said that after the library was carried to Rome, Tyrannio the grammarian arranged most of the works in it," reported Plutarch, "and that Andronicus the Rhodian was furnished by him [Tyrannio] with copies of them, and published them, and drew up the lists now current."

The key word there? *Arranged.* After the collection's long neglect, Tyrannio, recognized as an expert on Aristotle, restored order to the collection. He transformed a random jumble into a library with its own internal logic and meaning. Now he would do the same for Cicero.

When Tyrannio arrived, Cicero's library was in shambles, but he soon began sorting out the scrolls, identifying what was what, which volumes required repair, and so on. Cicero updated his friend Atticus on his progress and requested that he send Tyrannio some further assistance. "You will be surprised at Tyrannio's excellent arrangement in my library," he said. "What is left of it is much better than I expected: still I should be glad if you would send me two of your library slaves for Tyrannio to employ to glue pages together and assist in general, and would tell them to get some bits of parchment to make title-pieces, which I think you Greeks call 'sillybi.'"

As the scrolls were shelved lengthwise in wall cubbies, these title tags would hang from each volume and help a reader find the book he was after—like the title on a modern book spine. It's a small but essential way for the reader to engage the internal logic of the library. Individuated cubbies and title tags provided an interface that Cicero or any other reader could use to access the ideas on offer.

Tyrannio and Atticus's slaves soon finished the work, and Cicero wrote Atticus to express his appreciation: "Your men have beautified my library by binding the books and affixing title slips. Please thank them." He wasn't done being thrilled by the result. "Since Tyrannio has arranged my books," he wrote again, "the house seems to have acquired a soul. . . . Nothing could be more charming than those bookcases of yours now that the books are adorned with title slips."

A soul? Not quite. The Latin underlying the translation is *mens*, or mind. Another translation says "recovered its intelligence." And it's true; owing to both the ideas it houses and the logic by which it facilitates engagement, a library represents a sort of intelligence. Duke University classicist William A. Johnson draws this idea out by pointing to Cicero's use of the word when referring to the Divine Mind (*mente divina*) organizing the atoms of the world into the shapes we experience as creation.

And notice the comparison isn't to the product of the mind. He equates the library not with the physical world constructed *by* the intelligence but with the organizing principle or program that brought it about in the first place, the mind itself. Once organized, Cicero's library possessed a discernible—even if artificial—intelligence. It had a mind of its own. And that's true for all libraries.

Chapter 4

CLARITY AT SCALE—ALMOST

Look What I Found

Following a period of neglect and abuse, the Judahite temple in Jerusalem required some care and attention. King Josiah, who lived in the seventh century BCE, decided a makeover was due and dispatched his secretary to work out the details.

While the secretary sorted out the finances at the temple, the high priest dropped by with an announcement: "I have found the book of the law in the house of the LORD." No one can identify the book with total certainty, but scholars suppose it was part of Deuteronomy, the fifth book of the Pentateuch.

The king's secretary took and read. He then schlepped it back to the palace and read the book to the king, after which Josiah tore his clothes. Why the theatrics? According to the account in 2 Kings, Josiah lamented, "Great is the wrath of the LORD that is kindled against us, because our ancestors did not obey the words of this book, to do according to all that is written concerning us."

The king and his counselors evidently considered the book important and were concerned but also uncertain about its meaning. They wanted confirmation. So the high priest, the secretary, and a few others located a prophetess in town to "inquire of the LORD . . . concerning the words of this book that has been found." Sure enough, bad news: The prophetess said the curses detailed in the book would fall upon the king and the people.

Except, no! Right after her proclamation of doom, the prophetess said because the king deferred so humbly and penitently, God would likewise defer the book's judgment, at least for a while. The men returned to the king with the good news, and the account continues:

> *The king went up to the house of the LORD, and with him went all the people of Judah, all the inhabitants of Jerusalem, the priests, the prophets, and all the people, both small and great; he read in their hearing all the words of the book of the covenant that had been found in the house of the LORD. The king stood by the pillar and made a covenant before the LORD, to follow the LORD, keeping his commandments, his decrees, and his statutes, with all his heart and all his soul, to perform the words of this covenant that were written in this book. All the people joined in the covenant.*

It's a pivotal moment. "This was one of the most profound cultural revolutions in human history," says Ancient Near East scholar William Schniedewind. And what is "this" exactly? "The assertion of the ortho-

High priest Hilkiah shows King Josiah the Deuteronomic scroll. *SOURCE:* ART DIRECTORS & TRIP, ALAMY.

doxy of texts," he clarifies. Thanks to the z-axis, words written in times prior became binding on the present. More than recording history, the book now set its agenda. "This was the great and enduring legacy of the Josianic Reforms in the development of Western civilization," says Schniedewind, "the concept of textual authority." But, as we'll see, it wasn't as simple as the mere reading of words in a book.

Inescapable Tension

Despite the blessed delay, doom eventually descended. A dozen years after Josiah died, Babylonian king Nebuchadnezzar crashed the party, razed the temple, and hauled off Judah's best and brightest. Among those captives were the priests and scribes shouldering their sacred books all the way to the conqueror's far-off kingdom.

While temple and royal officials had worked to compile and edit the sacred books in their care before the invasion, the work continued in earnest during captivity. Prior to exile and destruction, the land and temple conferred Jewish identity. Now equally impermanent—though portable and copyable—books would have to perform that job.

There were other benefits to books as well. A book allows a single formulation of an idea to spread beyond an initial audience. It allows a message to scale with a community and even, as we'll see, beyond that community. People in different times and places can receive the same message. But there's a complication. It's right there in the Josiah story: Everyone could read the book, but they were unsure what it meant; they needed the prophetess.

The difficulty emerges even more visibly when the priest Ezra gathered the people in Jerusalem to hear the Torah sometime in the fifth century after two batches of returnees were free to come home. Standing on a dais, he "read . . . from early morning until midday, in the presence of the men and the women and those who could understand," according to the book of Nehemiah. Being able to understand and actually understanding are two different things. No worries—Ezra was ready. He positioned priests in the crowd to help explain and interpret as he went.

"They gave the sense, so that the people understood the reading," says Nehemiah. Without their aid, the people would have been lost.

It's an inescapable tension. Books promise clarity at scale but simultaneously elude comprehension. The book says what it says. But what, exactly, is that? Just as Josiah required the prophetess, the people surrounding Ezra required the priests. Books don't speak for themselves. To one degree or another, they all require interpretation and explanation.

You'll Need to Explain That

This need for explanation was baked into the use of books in Jewish synagogues. In Egypt, Syria, Greece, Macedonia, Italy, and beyond, Jews heard their scriptures not only intoned but also elucidated. Sabbath reading involved the whole community so Jews of all ranks could internalize the Torah.

Moses, said the first-century Alexandrian Jewish scholar Philo, "required them to assemble in the same place on these seventh days, and sitting together in a respectful and orderly manner hear the laws read so that none should be ignorant of them." Serving this purpose, "some priest who is present or one of the elders reads the holy laws to them and expounds them point by point till about the late afternoon, when they depart having gained both expert knowledge of the holy laws and considerable advance in piety."

Reserving one day in seven for worship and study peeved some people. Roman critic Juvenal mocked Jews for idling away to learn "all that Moses committed to his secret tome." But the scheme achieved its purpose. "Any one of them you attack with inquiries about their ancestral institutions can answer you readily and easily," said Philo.

Christians adapted synagogue practice for their own services. As Justin Martyr painted the scene sometime later, "On the day called Sunday, all who live in cities or in the country gather together to one place, and the memoirs of the apostles or the writings of the prophets are read, as long as time permits; then, when the reader has ceased, the president verbally instructs and exhorts to the imitation of these good things."

Whether at special public readings (such as Ezra's) or regular communal gatherings (like those described by Philo and Justin), explanation played—and continues to play—a key role in the life of books. In fact, the need to explain a book is an integral feature of the technology, which is ultimately responsible for several critical developments in the story of the idea machine.

What makes explanation integral? For one, thanks to the z-axis, some books last longer than the tongues they employ. Language evolves. Words fall out of use or find new purposes, changing meaning across generations. Thus the Hebrew scriptures, which were written over the course of a millennium, give or take, contain "different Hebrews" and "several distinct dialects," according to Harvard professor James Kugel.

English speakers can appreciate the problem by considering Shakespeare. Words, metaphors, and syntax accessible to his contemporaries stump us today. Even professionals struggle. "I haven't the faintest idea what they're talking about," admitted the director of Britain's National Theatre about lines spoken at the start of a play. He said it takes modern audiences ten or fifteen minutes to get up to speed. Even then, linguists such as John McWhorter insist that we miss more than we realize; we think a word means one thing when it really means another. As languages morph and modulate, they sometimes mislead. Kugel calls such words "false friends."

Then there are those who don't speak the language at all. As Palestinian Jews took Aramaic and diasporic Jews Greek for their primary tongues, Hebrew became more distant and undecipherable to many. Ezra's priests decoded for those who couldn't follow along, but needy ears outnumbered knowledgeable mouths. Translation eventually filled the gap and later, especially in Greek, secured the Bible's status as world literature.

Christians faced the same challenge. In his second-century polemic, *Against Heresies*, Irenaeus of Lyons spoke of bookless barbarians who had "salvation written in their hearts through the Spirit, without paper and ink." Like the converts themselves, literary virginity was foreign to the bishop's more civilized readers, for whom books were a regular feature of piety and worship. These Celtic newcomers lived "without writings" by necessity, not preference, as missionaries outpaced tutors and translators. In the meantime, audiences had to make do with whatever their preachers could manage to get across.

Translation comes with trade-offs of its own. Individual words—not to mention idioms and other modes of speaking—sometimes make ungainly leaps into new tongues. Cultural differences alone trouble comprehension. First-century Jews and pagan converts alike might struggle to understand the remote past of Lot's Sodom, David's court, or Ezekiel's visions. The greater the distance linguistically and culturally, the greater the need for help in understanding. Bishop Wulfia, missionary to the Goths, faced a version of this problem when he translated the scriptures into Gothic, choosing to leave out battle-filled bits like 1–2 Kings and 1–2 Chronicles to prevent his war-happy disciples from getting the wrong ideas. They were, as the historian Philostorgius says, "in more need of restraints to check their military passions than of spurs to urge them on to deeds of war."

Next, there are thorny, cryptic, and paradoxical passages to contend with. How should readers take the Genesis creation account, in which days somehow exist before the sun or moon? Or that God said Adam would die the same day he ate the forbidden fruit, and yet Adam woke the next morning and donned his fur britches before clearing thorns and planting vegetables? Ancient teachers offered different explanations for these puzzlers—and countless more—but explanations were needed nonetheless. As Augustine said in his *Confessions*, speaking to God before running through a string of possible interpretations in Genesis,

> *I have believed your scriptures,*
> *but those words are full of hidden meaning.*

Occasionally, the texts themselves indicate the difficulties they pose. The author of Peter's second letter says Paul's writings were sometimes "hard to understand" and that "unschooled and unstable" readers got him wrong. Likewise, the Gospels refer to "the abomination of desolation" (an unidentified allusion to the book of Daniel) and then add, "Let the reader understand." Only by looking *outside* the book could readers understand what was *inside*. Indeed, New Testament books like Romans, Hebrews, and Revelation required an ample grasp of the Hebrew scriptures to follow along—and even then perils persisted, contributing to an ongoing need for still more explaining.

It's impossible to encounter difficulties and discrepancies without attempting to reconcile them, whether by linguistics, textual comparison, allegory, or other hermeneutical gymnastics—either that or toss out the book, and that would never do.

Books are tools for explaining reality. People use myth, law, history, drama, and poetry to communicate something they regard as true about the world and how it works. What's more, the Jewish scriptures and later Christian additions claimed to impart the words of God, his dealings with people, and his ongoing expectations for their conduct: personally, communally, politically, and more. A lot was riding on the right reading.

Who's in Charge Here?

Orthodox Jews and Christians weren't the only producers of new books. Many in the classical world found the Christian story unbelievably odd. God born of a woman? Born in a Palestinian backwater? Rejected by his own people? Executed by a Roman governor? Hanged on a cross? Risen from the dead with the promise to raise the rest of humanity as well? "It would all have been less startling to the ancient mind," said historian Henry Chadwick, "if only the story could be cut free of its historical anchorage and interpreted as a cosmic or psychological myth attached to an esoteric mystery-cult."

Docetists and Gnostics were among the first to cut and reinterpret, producing in the process new gospels ascribed to Peter, Thomas, Mary, Judas, Philip, Matthias, and others. None had any direct connection to the figures in their titles; the names were chosen to lend credibility to their claims.

Authors appealed to tradition to validate their books. The events described by Luke, for instance, "were handed on to us by those who from the beginning were eyewitnesses." The evangelist wasn't there, but he spoke with those who were, which authenticated his account. The same was true for the Gospel of Mark, which was reportedly based on the testimony of Peter. But such apostolic links had more uses and acquired immense significance in response to the endless possible interpretations of existing books and the proliferation of new ones.

If books could be traced back to the apostles, so could their explanations, as well as the officeholders who did the explaining. Churchmen such as Irenaeus in the second century and Eusebius in the third and fourth traced their reading of the scriptures, along with what books counted as authoritative, back to the apostles through the succession of regional bishops—an office both Irenaeus and Eusebius held.

These leaders were seen as shepherds of the local flock and were charged, as Paul said, "to preach with sound doctrine and refute those who contradict it." At the turn of the second century, Clement of Rome claimed the apostles "appointed those we have already mentioned [bishops]; and afterwards they added a codicil [instructions], to the effect that if these should die, other approved men should succeed them in their ministry." Since later bishops were appointed by previous holders of the office, they served as links in a chain that extended back to the apostles themselves, which validated their views and their right to dispense them in the churches they oversaw.

Like readers today, literate Christians could consume and share whatever books they could get their hands on. One letter found in Egypt reveals a quid pro quo book swap. "To my dearest lady sister in the Lord," it begins, "greetings. Lend the Ezra, since I lent you the Little Genesis [the book of Jubilees]. Farewell from us in God."

Various gospels, testaments, acts, and apocalypses ascribed to biblical and apostolic figures were particularly attractive. It was up to the bishop to determine whether such books passed muster, especially for use in church. Irenaeus used his position as bishop to dismiss a gospel ascribed to Judas as a "fiction."

Around the same time, Christians in the northeastern Syrian city of Rhossus read a gospel ascribed to Peter. Their bishop thought nothing of it until hearing rumors that they had veered from orthodoxy. He procured a copy, read it for himself, and spotted a fraud. "We, my brothers, receive Peter and all the apostles as we receive Christ," he said, "but the writings falsely attributed to them we are experienced enough to reject, knowing that nothing of the sort has been handed down to us." The bishop wrote—what else?—a short book to expose the fraud, showing what portions of the gospel were authentic and which were "spurious additions."

In time bishops offered approved reading lists for the Christians in their charge, most famously the Alexandrian bishop Athanasius in his Festal Letter 39. Written in 367, it's the first that features the exact list of books in the current New Testament canon.

Two pages from a Gnostic codex, including part of the Secret Book of John, found in the Nag Hammadi Library. SOURCE: ZEV RADOVAN, ALAMY.

The process for deciding what ended up in Holy Writ (the four Gospels, Paul's letters, and so on) and what ended up in the rubbish bin (the Gnostic books) was the same. The authenticity and canonicity of the scriptures depended in large part on "ecclesiastical tradition" and "the succession of ecclesiastical writers," as Eusebius recorded.

The appeal to apostolic tradition allowed the early bishops of the church to define and defend the faith as they understood it. Books that failed to make the cut were in some cases destroyed, but most fell out of circulation and failed to be recopied.

Bishops oversaw the copying of books in their churches. Eusebius, for instance, kept a team of scribes to produce books in his church library at Caesarea. The institutional church, headed by bishops, was powerful enough to sustain copying efforts for the books they deemed valuable, whereas the fragmented Gnostic communities eventually died out, along with their books.

Socrates's Complaint, Again

Thinking of Josiah and Ezra's challenge takes us back to the problem Socrates raised in Plato's *Phaedrus*. Socrates complained a book was only able to articulate its message as formulated. Books, he griped, "just go on and on, forever giving the same single piece of information."

Books say what they say and can't take questions. The only way to clarify what's inside the book is to go outside the book. In more formal settings, that looks like appealing to the head of a reading community—priest, rabbi, bishop, pastor, scientist, professor, or someone of that sort. These authorities confirm the message or application of the book. In other words, there's what the book says and what they say it says, leaving room for rival authorities to vie for attention by complicating or contradicting those interpretations. The meaning is never fully settled.

Contrary to Socrates's view, however, this need for explanation is a feature of the technology, not a bug. The fact that a book can say one thing and we can read another in it means we can use the book to leap beyond the book.

Marginalia: By the Book

Hidden in the gleam of the sun, an unknown "worthy soul" awaits Dante and Beatrice as they ascend to the second heaven in Cantos 5 and 6 of Dante Alighieri's *Paradiso*. "If you desire enlightenment," says the radiant being, "ask to your heart's content." Dante wonders about its identity.

> *Caesar I was, Justinian I remain*
> *who, by the will of the First Love I feel,*
> *purged all the laws of excess and of shame.*

The Italian word here translated as "shame" is *vano* (related to the English term *vain*), meaning he cut whatever was unnecessary, unhelpful, and inefficient from Rome's laws. And, yes, that's right: Extending hope to all of us poor souls who toil with words, Emperor Justinian made it to heaven because he was a good editor.

The memory of all great lawgivers—think Hammurabi, Moses, Solon, Ezra—is inextricably linked to writing. Referencing our idea grid, written laws embody the human attempt to move an idea along all three axes:

- the x-axis, publishing the idea so it can be known by all to whom it applies;
- the y-axis, rendering its meaning explicit, plain, and unambiguous; and
- the z-axis, preserving the formulation through time so it can be referred to and appealed to at later dates.

In the early Roman republic, the patrician class abused the prevailing legal structure by arbitrarily wielding unwritten customs as law over the plebeians. To fix this problem, in 451–450 BCE authorities instituted the Law of the Twelve Tables, which applied equally to all and covered everything from family law to property, debt, personal injury, court procedures, punishment, and more. A few gems:

- "Whoever is convicted of speaking false witness shall be flung from the Tarpeian Rock."
- "Slaves caught in the act of theft . . . shall be whipped with scourges and shall be thrown from the rock."
- "If anyone sings or composes an incantation [song] that can cause dishonor or disgrace to another . . . he shall suffer a capital penalty" (that is, be clubbed to death).

People were stricter then. But however precise the laws might have been, they couldn't overcome all ambiguity, or anticipate every interpretation, or cover every circumstance. Back down the *y*-axis they slid. So, over the centuries, all manner of official qualifications, rulings, and additions encrusted themselves on the law to clarify and extend it. But none of these supplements and modifications were compiled in one place for easy consultation, making it practically impossible for lawyers, judges, and others to know what the law really said. Down the *x*-axis they slid as well.

The solution? The first step was to gather all these scattered edicts and statutes in one collection. Around the time when Emperor Diocletian assumed power in 284 CE, an anonymous editor compiled the *Codex Gregorianus*, which arranged imperial legislation in some fourteen volumes by title and topic for use by lawyers and other legal professionals. But it wasn't exhaustive. Nor was the *Codex Hermogenianus*, which later collected Diocletian's legislation. In 429, Emperor Theodosius sponsored his own collection, and various barbarian kings in the Western Roman ambit followed suit.

The end result was confusion. Though collections existed for consultation, they were rife with contradiction and superfluous and outdated material. On top of that, none addressed all the free-floating records and rulings of jurists whose opinions had the legal authority of case law but weren't available in any sort of systematic way, despite totaling 2.4 million to 3 million lines of Latin and running to somewhere between sixteen hundred and two thousand individual volumes.

In stepped Emperor Justinian in 529 with an ambitious plan to clean up the mess. He started by commissioning the *Codex Iustinianus*, a collection of imperial legislation going back four centuries, stripped down, streamlined to the essentials, and revised for the times. Crucially, the entire collection would fit between two covers. He next had the prolific writings of the jurists compiled and radically abridged to fit a single volume as well (the *Digest*); the same applied for a new legal manual (the *Institutes*). He then had the original collection updated. "Thus," the emperor told the Senate, "the entire assemblage of Roman law has been compiled, completed in three volumes. . . . We have given ample thanks to the Supreme Deity, who has enabled us . . . to lay down the best laws not only for Our own but for every age, both present and future." Here we see Justinian's concern for the *x*-, *y*-, and *z*-axes.

Indeed, Justinian's codification of Roman law long outlived the Roman Empire. It remained operative in limited parts of the Italian peninsula but, once embedded in Catholic canon law, spread throughout Europe in the Middle Ages. Other aspects of the collection resurfaced in the eleventh century and began to influence legal education and practice, forming the basis of European common law and remaining influential well into the nineteenth century. Dante's world relied on Justinian's work; so did Erasmus's and Voltaire's.

Human societies existed for millennia before the advent of writing, making do with orally transmitted rules and norms, as the Romans had done before the Twelve Tables. However, as the Romans discovered, complex social arrangements were impossible without writing and clear, objective standards to which people might appeal when necessary. But the technique of writing wasn't enough. For the law to keep pace with societal evolution, the technology of the book required updating as well, and we can see the innovation in the compilations just mentioned.

Scrolls were the most common book form in the ancient world, but the *Codex Gregorianus*, *Codex Hermogenianus*, *Codex Theodosianus*, and *Codex Iustinianus* represented something new. We use words like *code* or *codify* today in legal contexts without thinking of where these words come from. They refer not to the laws themselves but to the book format in which the laws were recorded.

A codex is a book of pages bound at the spine, still something of a novelty during these centuries of compilation and reform. We'll give more attention to the codex in the coming chapters, but for now it's enough to say that what Justinian accomplished would have proven technologically impossible were he dependent on scrolls. As it stands, the adoption of the codex changed the entire shape of Western legal thought and practice, with effects that not only lingered into the modern era but also helped produce it.

Chapter 5

NO BOOKS FOR YOU

Cicero and Atticus, Pen Pals

In the summer of 59 BCE, a man named Vibius loaned Roman states-man Cicero some poetry, a volume by Alexander Lychnus of Ephesus. He didn't enjoy it. "He is a wretched poet and indeed has nothing in him," Cicero wrote his friend Atticus, adding, "still he is of some use to me. I am going to copy the work out and send it back." Sure enough, a few days later Cicero sent an update: "I have sent back the works of Alexander, who is a careless writer and not much of a poet: still there is some use in him."

Cicero's letters to Atticus highlight at least a couple of underappreciated aspects of book culture in the ancient world—back when Amazon referred not to an online store but to a tribe of exceptionally dangerous women by the Black Sea. In these two letters, for instance, we see a primary way books circulated: through networks of friendly lending and borrowing. And what if the recipient enjoyed the book enough to want a copy—or at least found "some use" in it? They could make a copy them-selves, as Cicero seems to indicate. They might at least make extracts like Sidonius's friend, as we learned earlier.

But copying books was tedious work, and Cicero probably didn't have the time or the penmanship for that. Instead, depending on one's means, a person might have a dealer secure them a copy or hire a scribe to reproduce it for them. If they were wealthy or possessed ample

Cicero, probably jotting a missive to Atticus. *SOURCE:* IANDAGNALL COMPUTING, ALAMY.

business interest to justify the expense, they'd have a slave do it. Atticus was a book dealer and owned several scribes, and Cicero was wealthy enough to count book-copying slaves as part of his household, at least for a time; he often resorted to Atticus's services in periods when he possessed no slaves of his own.

Considering these two letters, we find ourselves wandering through radically different territory than readers traverse today, one marked by a kind of literary scarcity we can hardly imagine. Ancient manuscripts alone tell us as much: Glance over these artifacts, and certain details conspire to tell a story. I'm not talking about epic tales of mythic heroes or the gripping tit-for-tat of shrewd philosophers. No, despite the remarkable achievements of the classical world, viewed from today's vantage point, their manuscripts tell a story of limitations, even deprivations.

While we rarely think of it, the literature and learning of the ancients were inaccessible to almost all ancient people. Literacy was rare, as were books, and class limitations barred the majority from intellectual pursuits.

Books belonged to the elites, people such as Cicero, Atticus, and the names they drop in their letters. Given innumerable, inexpensive copies of Aristotle, Seneca, Livy, and—of course—Cicero available on Amazon today, it's easy to overlook the practical, cultural, and economic barriers that prevented contemporaries from reading their words.

Start with the simple fact that books are physical objects. Since books involve ideas, we sometimes forget they're made of stuff, and what that stuff is and how it's used has ramifications.

Physical Limitations

Before the mass production of paper, writing materials were hard to come by. Bark, leaves, clay, and cloth all served in one time or place. Many of our words for books come from the materials they were written on. The Latin *liber* (think library) first referred to tree bark. The Greek *biblion* (think Bible or bibliography) stems from the city of Byblos, a major papyrus exporter. This is true for English, too; *book* comes from the old Germanic word for beech.

The Latin codex originally denoted a block of wood. As we saw earlier, Greeks, Romans, and others filled recessed wooden tablets with beeswax and scribbled notes on the reusable surface with a stylus. Wordy users bound several tablets along one side and created bulky notebooks (codices) for keeping accounts, drafting documents, and other workaday purposes. (More on the evolution of the codex in a bit.)

For more final or formal uses than drafts and accounts, there were papyrus and parchment. Papyrus was made from reeds native to the Nile and marshes of Egypt. Workers harvested the triangular stalks, peeled them, and cut their pith in lengthwise strips. They moistened, layered, and pressed these into sheets that were later smoothed for writing. Glued end to end, the sheets were rolled up and sold by quality and length. A roll could cost as much as a week's pay for a laborer but was affordable for the wealthy and powerful, who used papyrus most.

Distance from the source and the health of the crop affected price and supply. So did politics. In his *Natural History*, Roman historian Pliny the Elder tells of a tiff between the monarchs of Egypt and Pergamum over who possessed the better book collection. Who wore it better, the famed library of Alexandria or that of Pergamum? Supposedly, the Ptolemaic king banned papyrus exports to prevent his rival from producing more volumes. Some doubt the story and blame war for interrupted supplies. Either way, Pergamum found an excellent work-around in parchment.

Parchment makers stripped the skin from an animal—usually a calf, goat, or sheep—and soaked it in a lime solution. After sufficient scrubbing, stretching, scraping, drying, and polishing, parchment makers produced a thin, pliable sheet that took ink as well as papyrus did, possibly better. It was also durable. Pliny praised it as "the material on which the immortality of human beings depends." Best of all, cows were everywhere, as were sheep and goats. Freeing scribes from the geographical, agricultural, and political limitations of a local weed, the use of parchment "spread indiscriminately," said Pliny.

But parchment possessed limitations of its own. Preparing hides was laborious, and a single book could require many animals—the tally rising with the length of the tome. As a result, supply for both papyrus and parchment could scale only so far. That capped the availability of books and helps explain why two monarchs might compete over their collections. But the scarcity was just as well; few people could read books anyway.

Technical Limitations

Take another look at those manuscripts. While spacious margins frame neat, narrow columns of text, lines appear without space between words or punctuation, at least in Greek and Latin books. Readers might add their own punctuation—students parsing knotty passages, for instance, or orators preparing for a public reading. But scribes themselves made virtually no use of it. What met the eye resembles a "river of letters" or a "monolith of characters," as two scholars variously put it.

Readers of this *scripta continua* style, which the Romans adopted from the Greeks, didn't so much identify words as pick out syllables and arrange them into coherent combinations, slowly. Why slowly? Well, try finding your way to the bottom of this cluster:

TELLMEOMUSEOFTHEMANOFMANYDEVIC
ESWHOWANDEREDFULLMANYWAYSAFTERH
EHADSACKEDTHESACREDCITADELOFTROYMA
NYWERETHEMENWHOSECITIESHESAWAN
DWHOSEMINDHELEARNEDAYEANDMANYTH
EWOESHESUFFEREDINHISHEARTUPONTHESE
ASEEKINGTOWINHISOWNLIFEANDTHERETU
RNOFHISCOMRADESYETEVENSOHESAVEDNO
THISCOMRADESTHOUGHHEDESIREDITSOREF
ORTHROUGHTHEIROWNBLINDFOLLYTHEYPE
RISHEDFOOLSWHODEVOUREDTHEKINEOFHE
LIOSHYPERIONBUTHETOOKFROMTHEMTHED
AYOFTHEIRRETURNINGOFTHESETHINGSGODDE
SSDAUGHTEROFZEUSBEGINNINGWHERETHOUW
ILTTELLTHOUEVENUNTOUS

So begins Homer's *Odyssey* in A. T. Murray's old translation: "Tell me, O Muse, of the man of many devices, who wandered full many ways after he had sacked the sacred citadel of Troy." Someone familiar with the format could read faster and more reliably than we might, but rappelling down the face of an impenetrable column of text was hardly speedy even for the most adept.

Discrete words help us rapidly decipher word after word, line after line. Without spaces, our eyes move unevenly, first darting, then slowing, trying to sort and assemble the syllables as we go. Meanwhile, our short-term memories pile up possible meanings until we arrive at a complete thought. It's taxing work. Word spacing and punctuation shoulder much of the burden for us today, which accelerates decoding and amplifies comprehension. Classical educators compensated by forcing students to memorize every conceivable syllabic combination. Further compounding the challenge, classical Latin composition often relied on cumbersome, nonintuitive sentence structures. Good luck!

These features made scripta continua texts inherently ambiguous and tough to understand. Roman grammarian Aulus Gellius illustrated the difficulty in *Attic Nights*. After encountering a man trumpeting his interpretive skills, Gellius surprised him with a book the braggart supposedly knew forward and backward. But no. The man fumbled from the start. "Ignorant schoolboys, if they had taken up that book, could not have read more laughably," said Gellius, "so wretchedly did he pronounce the words and murder the thought." As onlookers laughed, the "egregious blockhead" returned the book and blamed tired eyes for his poor performance. Despite his vaunted expertise, the man couldn't scale the monolith.

Authors skirted such difficulties by sharing their books at public readings, revealing firsthand their own sense of how phrases fit and flowed. A *recitatio* (as such an event was known) was among the most common ways to discover and experience literature in the ancient world. Listeners could hear the pauses and inflections, essential for understanding but absent from the bare text. On their own, however, readers were left to their own devices.

Today almost everyone can read. Literacy is near universal in developed countries. Almost nine in ten adults around the globe qualify as literate, according to the United Nations. Among more than two hundred countries, literacy dips below 50 percent in just twenty. But that's now. Even with Rome's extensive school system, literacy in the classical world rarely exceeded 20 percent of the population. In special cases, it might rise to as high as a third, but, according to William Harris's *Ancient Literacy*, at the fringes of the Roman Empire it dropped to as low as 5 or 10 percent. And it's safe to say that's how they wanted it.

Social Limitations

Classical society was highly tiered. Tradespeople might have enough skill to read graffiti or brief documents, but they were at a loss with longer, more complicated literary or philosophical texts. Literary pursuits reinforced deliberate and (for the elite) desirable inequality. "The ancient world did not possess the desire, characteristic of the modern age, to make reading easier and swifter," says Paul Saenger of Chicago's Newberry Library. "The notion that the greater portion of the population should be autonomous and self-motivated readers was entirely foreign to the elitist literate mentality of the ancient world."

As an icon, the book represented status. Scaling the monolith demonstrated attainment, cultivation, and taste. Books in the arduous scripta continua style "were designed for clarity and beauty, but not for ease of use, much less for mass readership," according to Duke University classicist William Johnson. When it comes to "works written expressly for the masses," says Harris, "there were none."

Far from serving as tools for self-advancement, books reinforced the barrier to entry for anyone outside the ranks of wealth and status. Learning was reserved for the upper class; note that our word *school* comes from the Greek *skhole*, which refers to leisure (a ready point of disagreement for modern high school students). Literary culture was underwritten by the privileged for the privileged.

Of course, reading was still work even for those who knew how. For elites who tired of the tedium, slaves were trained for the task. Pliny the Elder kept slaves to read to him when he was bathing, being carted around the city, or reclining outdoors on summer afternoons. Slaves were

Pliny the Elder at rest while two attendants (likely slaves) read behind him.
SOURCE: CHRONICLE, ALAMY.

plentiful in the ancient world, constituting as much as a third of the population in the Italian peninsula; in Athens, they outnumbered citizens two-to-one. It took an economy supported by mass bondage to provide the upper crust with free time to read.

As seen with Cicero and Atticus, literate slaves were also useful in the production and reproduction of books. If reading could be tedious, writing was worse. As the grousing of scribes in colophons and marginalia attests, copying manuscripts was a chore. The second-century medical doctor and writer Galen mentioned "the writings of the ancients that I had copied by my own hand," but he was an outlier. When Cicero said he would make a copy, he likely meant he'd have a slave do it. This point is further underscored by the fact that the ancients denigrated manual labor, which manuscript copying most definitely was.

The majority who could read and write placed no value on the acts of reading or writing—any more than they did on herding pigs. As a result, manuscript production relied on slaves or freedmen (those once enslaved but later freed). The use of slaves was so ubiquitous that writer Rex Winsbury referred to them as the "infrastructure" of the trade. Atticus's business as a book dealer depended on his team of in-house slaves who could produce copies for those who requested them.

Economic Limitations

Under such conditions, it's impossible to speak of a "publishing industry" or "book trade," at least not in the terms and on the scales to which we're accustomed. Not only were books not published for the masses—reading being the preserve of the privileged—but doing so was also physically impossible.

Today a digital file is sent to a printer who reproduces thousands of exact copies, binds them, and ships them to distributors and retailers. Or a version of that digital file is uploaded to retailers, who make it available for customers to download. In either case, readers—potentially hundreds of thousands of them—all possess or access the same text. But in a manuscript world, every copy of a book was a one-off. No two copies were exactly alike, and mass production was impossible, unthinkable. Only a few authors whose works ended up as the basis of schooling—say, Homer or Vergil—experienced what we might consider mass readership.

So how did people find and acquire books? In his study of Roman libraries, classicist George W. Houston offers about a dozen different means by which readers built collections. We've already mentioned a few. A person might copy a book themselves (unlikely) or have a slave do it (more likely). Books might change hands as part of a bequest. The books of Servius Claudius, a man in Cicero's social circle, found their way to Cicero under such circumstances.

Cicero wanted Atticus's hand in getting Servius's books from point A to point B: "Now, as you love me, as you know I love you, stir up all your friends, clients, guests, freedmen, nay even your slaves, to see that not a leaf is lost. For I have urgent necessity for the Greek works, which I suspect, and the Latin books, which I am sure, he left," adding in a follow-up, "This gift depends on your kind services. As you love me, see that they are preserved and brought to me. You could do me no greater favour: and I should like the Latin books kept as well as the Greek. I shall count them a present from yourself."

An influx of books like that into one's library was a windfall and didn't happen often. Other means were equally rare: Sometimes libraries were seized for legal reasons or taken as war booty, their contents suddenly available for acquisition by whoever possessed proximity to power.

To mention one example we saw earlier, when he sacked Athens in 86 BCE, the Roman general Sulla seized the library of Apellicon of Teos, a prominent book collector in the city. Apellicon relied on another of Houston's dozen methods to build his own collection of rare books and documents. I'm referring not to theft, to which Apellicon sometimes resorted, but rather to the purchase of existing collections—most famously the books of Aristotle and Theophrastus, which he bought from the descendants of Neleus, who had originally inherited them. As we also saw, it's to that purchase and Sulla's subsequent capture that we owe (at least in part) the spread of Aristotle's ideas, not to mention those of other Greek philosophers, in the Roman world.

After the assassination of Julius Caesar in 44 BCE, Mark Antony added public intellectual Marcus Terentius Varro to his list of enemies.

Officials confiscated Varro's property, including his massive library, with some 490 volumes he had authored himself. "We cannot be sure what happened to Varro's manuscripts," says Houston, "but it seems very likely that some or all of them were seized by Mark Antony, passed on Antony's death to his wife Octavia, and were given by her to the library she helped found in the Portico of Octavia." Some years before, in 58 BCE, Cicero himself had been exiled and at least part of his library seized or otherwise damaged, as we saw earlier.

It's safe to say there's nothing scalable about this way of building a library. It doesn't increase the overall number of books but merely transfers them from one possessor to another. At the end of the day, the only way to get more books was to have them reproduced, one hard-won copy at a time.

Libraries and bookshops did exist, especially in urban areas, and enterprising booksellers found ways of ramping up production. The poet Martial mentioned dealers in Rome who peddled his pages. "Seek out Secundus, the freedman of learned Lucensis, behind the entrance to the temple of Peace and the Forum of Pallas," he advised. He also recommended another shop "with its doorposts from top to bottom bearing advertisements, so that you can in a moment read through the list of poets. Look for me."

But a dealer could produce only so much inventory on speculation in advance of real customers. Though Martial said this dealer stocked his books in attractive purple covers, shops held little in the way of actual inventory. Most volumes on hand were exemplars. When a customer wanted a book, the owner passed the exemplar to his slave, who then scratched out a copy for the customer.

Commonly, as Houston points out, books were privately produced within social networks. Authors would have copies made for a patron or friends, and readers loaned their books for others to read and copy, as Vibius did with Cicero. Atticus would sometimes send Cicero books he suspected his friend might enjoy. "I am much obliged to you for sending me Serapio's book," wrote Cicero, "though between you and me I hardly understand a thousandth part of it." While the act was generous, it wasn't a gift as such; Cicero was keen to pay his friend for the volume: "I have given orders for you to be paid ready money for it, to prevent your entering it among presentation copies."

Within these networks, access to books depended on who knew whom. Cicero's friends availed themselves of his close relationship with Atticus. An example: "Thyillus asks you, or rather has got me to ask you, for some books on the ritual of the Eumolpidae," Cicero wrote Atticus on behalf of a friend.

It was rare for a book to circulate outside a narrow network, even if authors wanted their work to spread. And this social sharing of scarce books points to another important limitation that positively impacts the rest of our story.

The Breakthrough Nobody Noticed

The narrow circulation of books and the difficulty of their manufacture created the necessity for users to read with eyes and ears toward useful takeaways. A reader would sometimes borrow a book to copy in full for his library. But, as we saw earlier, he might also make extracts, jotting down whatever passages he deemed most interesting or valuable.

As Pliny's nephew remembered of his uncle, "He would frequently in the summer, if he was disengaged from business, lie down and bask in the sun; during which time some author was read to him, while he took notes and made extracts, for every book he read he made extracts." In the preface to his *Natural History*, Pliny says he sifted through two thousand books for the facts he found valuable enough to share. "No book was so bad but some good might be got out of it," he famously said.

The first-century Roman philosopher Seneca advised readers to mimic bees, who gather nectar across wide fields of flowers, pack it away, and then transform it into their own creation. "We also," he said, "ought to copy the bees, and sift whatever we have gathered from a varied course of reading, for such things are better preserved if they are kept separate; then, by applying the supervising care with which our nature has endowed us . . . we could so blend those several flavors into one delicious compound that, even though it betrays its origin, yet it nevertheless is clearly a different thing from that whence it came."

This sort of reading engaged a latent critical function in the reader. With limited access to books and scarce resources for their reproduction and retrieval, readers were constrained to judge what was worth the time and trouble to record, necessarily reflecting the particular vantage and values of the individual reader. Seneca speaks to this process directly: "Reading," he said, "enable[s] me to pass judgment on their discoveries and reflect upon the discoveries that remain to be made."

Critical engagement with books enlivened the mind as it does today. Over a lifetime of reading, a collector might amass quite a stack of useful extracts. Pliny left his nephew 160 commonplace books "in a small hand" full of his reading notes. These were small scrolls, written on both sides, but by the late first century a more advantageous format was emerging to store all this choice nectar.

Waxen wood codices were best suited for temporary notes and figures. They were easy to erase and reuse, which is why poets found them desirable for early drafts and storekeepers employed them for orders and tallies. But eventually someone realized they could stack, fold, and bind parchment sheets as easily as wood tablets. In fact, Julius Caesar might have done it first. These codices of parchment—called *membranae* from the Latin for skins—were excellent for more enduring work. Depending on preference and availability, papyrus codices also served, though perhaps not as durably.

Jotting extracts, observations, and judgments in a parchment or papyrus codex created a way for readers to leaf through past thoughts and reengage old questions with fresh eyes—all the more when the extracts were arranged by topic, which was commonly done. Unlike scrolls meant for reading start to finish, codices were meant for thumbing, rummaging, and easy reference. University of Reading historian Matthew Nicholls notes that users sometimes even created personalized indices to more quickly locate useful passages. These were books "designed for dipping into," he says, and they could be extremely valuable, depending on their contents.

Galen used codices to store pharmacological research. He bragged that his manuals held "more remarkable drug recipes" than any others around. Easy access to a range of cures allowed Galen to ably serve his patients and quickly find recipes for trade when another physician had a concoction he desired.

Some authors and booksellers advocated the parchment codex for literary works, but it was a hard sell for a society used to the scroll. Martial upheld the format, praising its relatively smaller size compared to scrolls. "You, who wish my poems should be everywhere with you, and look to have them as companions on a long journey, buy these which the parchment confines in small pages," he said. "Assign your book-boxes to the great, this copy of me one hand can grasp." Martial also spoke of his "little book" being "thumbed everywhere," a verb that works only with the codex. More than a self-promoter, Martial plugged Homer, Vergil, Cicero, and others in the same format.

Marketing has long posed one of publishing's toughest challenges—something the poet soon discovered. Nobody listened! Statistical analysis by University of Edinburgh historian Larry Hurtado reveals that only a tiny number of classical books in this period ever found their way into codices. Consider Homer, one of the most widely published authors of the classical world. Of the nearly eight hundred second-century copies of his work found in the Leuven Database of Ancient Books, only 2 percent are codices.

There was much to recommend the codex. Winding and unwinding a scroll for a certain thought or story couldn't compare with flipping pages. Back to Augustine in the garden: Readers of the new format could access any part of a book on a whim, allowing heightened, more inventive, and personal engagement with its words and ideas. One scholar compared its importance to the invention of the wheel! But few appreciated the breakthrough at the time.

So how did the new format catch on? Not by reading Homer, Vergil, and Cicero—let alone Martial. To answer the question, we must return to Palestine. While the parchment codex originated in the West, its sectarian use in the East changed the trajectory of the book and the societies it transformed. But first, let's sum up what we've explored thus far.

Marginalia: Augustine's Tome

Alaric's message was simple: Treat your friends better than your enemies. The Romans, not quick to listen to barbarians, ignored the message of their bellicose Gothic ally and in 410 CE found Alaric's hordes on the wrong side of the city walls, running through the streets, plundering as they went.

News of the sack sent shockwaves through the Roman Empire. Word traveled faster than the refugees who fled the city looking for safer environs, the masses distraught and puzzled by the development. The empire's administrative capital had moved to Ravenna, but what did the fall of Rome, the supposed Eternal City, mean? It presented an existential crisis.

Alaric didn't stay long, but the questions unleashed by his invasion lingered and multiplied. How did it happen? Who was to blame? Rome was now in the midst of an epochal transition, its religious allegiances to the old gods slowly and somewhat grudgingly giving way to Christianity. But wasn't this calamity proof that the Christian God was worthless? He had obviously failed to protect the city as the old gods had done successfully for eight centuries.

Across the Mediterranean, Augustine heard the complaints in Hippo. "I decided to write the books, *On the City of God*, in opposition to their blasphemies and errors," he later recalled.

He'd been thinking about these and related issues for at least a year before the sack of Rome. Barbarians had been causing trouble in other parts of the empire as well, especially in Italy, Gaul, Spain, and Egypt. In 409, a priest asked him whether he'd write "an extended work" on the problem. His letter in response contains ideas that would work their way into *The City of God*, but he saw no reason to tackle such a project at that point. Other aspects of *The City of God*, particularly the opposition of the heavenly and earthly cities and their respective allegiances, had been percolating in his mind for years, emerging in his other writings going back a couple of decades before he finally began working on this, his most complete statement.

While Augustine responded to the sack in sermons preached at the time—probably with refugees standing in the congregation—he waited two years to begin writing. As a bishop, he had other questions and controversies to deal with. But with the various strands of thought already in play, it seems likely he spent the periods between other obligations both actively and idly noodling on how they fit together, and the longer he thought, the greater the project grew.

Augustine ended up writing twenty-two books on the subject, what we would likely recognize as large chapters or sections of a single book, tackling questions of history, religion, politics, and philosophy, all triggered by the immediate crisis but extending into considerations ranging far beyond. The scope of the work took extensive conceptual effort and intense cognitive labor, requiring some fourteen years to ultimately complete. In many ways, the resultant book exemplifies what the technology itself made—and makes—possible.

Consider the two primary themes of the opening section of this book: thought and communication. Augustine had a big idea, but it was at first fuzzy and fragmentary. He'd written bits and pieces of that idea, but they were underdeveloped and didn't cohere as a unified statement. Writing allowed him to develop and systematize his ideas, working and reworking the various strands and tangents as he went.

Augustine wasn't working with memory alone, or even with personal thoughts he could build on iteratively as he progressed from one page to another. He had other pages to consult, as well as scrolls. Augustine's book is, perhaps more than any book before it, a book of other books. Through his interaction with his own prior thinking and the authoritative texts of the Greco-Roman world, his fully fleshed theory came into view.

Thoughts progress by rebuttal. We learn by contradiction and argument. But how could Augustine buttonhole Vergil and Varro, Cicero and Seneca, Terence, Sallust, Livy, Lucan, Pliny the Elder, Aulus Gellius, Apuleius, Plato, Plotinus, and Porphyry—not to mention Jewish and Christian authorities such as Josephus, Philo, Origen, Eusebius, Jerome, Orosius, and others? Most were dead or otherwise indisposed. But they all had books, and, to repeat an earlier quote from economist Tyler Cowen, "once an idea has been generated, it can be used many times by many different people at very low marginal cost." As the words in the garden could bring Augustine solace, these words could provoke realization, disagreement, and deeper engagement to reappraise, refute, and reformulate. "The Rome he deconstructs," says Gillian Clark of Augustine's work in *The City of God*, "is made of books and his commentary on those books."

All the while, the effort forced Augustine to move his thinking higher up the *y*-axis, as he worked to state and restate his views in their clearest, most complete formulation. This process took time, not only for the iterative process of writing but also for feedback from early readers and the book's further development.

As the book moved up the *y*-axis, it also moved along the *x*-axis, being fully expressed as Augustine published it in invitational readings and through sharing the manuscript among his network, letting it out to be copied. What's more, he could trust that the same process that allowed him to encode such a sweeping, complex, intricately constructed, closely argued statement to begin with would also allow readers to decode what he intended (albeit with some inescapable ambiguity inherent to all human communication).

Just as he was forced to revisit earlier passages as he worked on later passages, ensuring that he provided adequate proof for his claims and tied up loose ends, so his readers could flip through the pages to take up one point or another to ensure that they understood the issue under consideration. And while the passage of time and the evolution of cultures might make some of his ideas difficult to comprehend today, the book is with us today thanks to its travel along fifteen hundred years of the *z*-axis, affecting readers during every century between then and now. But this? It's only the beginning of our story.

DEVELOPMENTS

The greatest benefactors of mankind are unsung and unknown—the inventor of the wheel, the deviser of the alphabet. Among their number we should place the inventor of the codex.

—ERIC G. TURNER

Humans don't decide what to build by making choices from some cosmic catalog of options given in advance; instead, by creating new technologies, we rewrite the plan of the world.

—PETER THIEL

Chapter 6

TURNING A NEW PAGE

On the Road

After the ascension of Augustus in 27 BCE, Roman roads and shipping lanes became relatively safe, even speedy. The cities of Mediterranean Europe, North Africa, Asia Minor, and the Levant were reachable by foot or ship in a matter of days and weeks. Traveling from Rome to Antioch took a month and a half, depending on the season, route, and conveyance. A person could walk from Jerusalem to Alexandria in about three weeks. Mere days separated Corinth, Thessalonica, and Ephesus.

Nobody exemplifies the ancient traveler better than St. Paul. Despite references in his letters to delays and shipwrecks, he was impressively mobile, crisscrossing the eastern side of the Roman world several times over. But not even a road warrior like the apostle could be everywhere. Fortunately for his fledgling movement, the empire's ample lanes accommodated messengers as well as missionaries.

Christians turned this unprecedented connectivity to their own purposes—exchanging letters, stories, and treatises. Jesus of Nazareth left us nary a syllable; yet his followers appear among the wordiest in the ancient world. Imagine authors dictating and scribes jotting shorthand on wax tablets, finalizing drafts on papyrus or parchment. Then picture couriers running texts this way and that to be copied and further disseminated. One scholar called this buzzing network the "holy internet."

Paul mentioned this literary networking in his Epistle to the Colossians. "And when this letter has been read among you," he said, "have it read also in the church of the Laodiceans; and see that you read also the letter from Laodicea." Most of his letters made the rounds this way. Though the letter to the Laodiceans is no longer extant, the rest make up a significant amount of the most enduring artifact of the young faith, a book we call the New Testament.

The apostle supposedly came off as more forceful on the page than in person. "His letters are weighty and strong, but his bodily presence is weak, and his speech contemptible," he quotes his detractors as saying. Despite those critics, Paul's letters were widely read and deemed authoritative early on. The author of the Second Epistle of Peter assumed broad familiarity with them, indicating some sort of collection to which far-flung churches had access. These were the same letters in which Augustine found solace under the fig tree.

But it wasn't just Paul's letters. Christian literature of all sorts circulated on the holy internet.

Traffic Report

Take the second-century book *The Shepherd of Hermas*. Approached by an elderly woman personifying the church, Hermas is told to make two copies of a small book, one of which is meant for Clement of Rome, "who will then send it to the cities abroad, for that is his duty." Clement apparently did his job. As Cornell professor Kim Haines-Eitzen says in *Guardians of Letters*, "soon after" the book's composition in Rome, "we find papyrus copies of the text in upper Egypt, quotations in Clement of Alexandria and Irenaeus of Lyons, and some forty years later a mention in the Muratorian Canon." All of this demonstrates widespread copying, sharing, and use.

Haines-Eitzen points to a similar pattern for the account of Polycarp's martyrdom. A disciple of the apostle John, Polycarp was bishop of Smyrna in the mid-second century. After his execution by fire, members of his church penned the story, sent it to another church in their network,

St. Paul with—what else?—a book. The codex hadn't developed this far yet. But the apostle and his followers would ensure that the world adopted their favored book format even more thoroughly than their faith. *SOURCE:* GOOGLE ART PROJECT, WIKIMEDIA COMMONS.

and instructed its members to digest and spread the word: "When you have heard these things, send a letter to the brethren further on." Irenaeus of Lyons had studied under Polycarp; he ended up with a copy, which was then further copied, scribe after scribe detailing the transmission in an evolving postscript: "This account Gaius copied from the writings of Irenaeus . . . and Isocrates used in Corinth the copy of Gaius. And again I, Pionius, wrote from the copies of Isocrates . . . gathering them together when they were almost worn out from age."

Polycarp himself bundled and shared letters, most famously the letters of Ignatius, bishop of Antioch. On his way to Rome in chains, Ignatius wrote to churches in six different cities. After his captors changed their travel plans and cut short his correspondence, Ignatius directed Polycarp to write additional letters on his behalf "because," he said, "I have not been able to write to all the churches. . . . Some can send messengers by foot; but others can send letters through those whom you send yourself."

When Ignatius reached Philippi, locals there heard about his letters and wrote to Polycarp, asking him to send any he possessed. "We have forwarded to you the letters of Ignatius that he sent to us, along with all the others we had with us, just as you directed us to do," Polycarp answered in his cover letter to the collection. "You will be able to profit greatly from them." The time between the request and its fulfillment was probably just a few weeks, according to University of Virginia professor Harry Gamble. The complete collection soon became a literary unit, like the Pauline letters, and was shared far and wide.

"In all this," says Gamble, "we glimpse a busy, almost hectic traffic of messengers and letters between churches of the region." These exchange networks reveal an appetite for news, encouragement, and instruction on the part of senders and recipients. These elements form the core of Paul's correspondence, along with that of Ignatius, Clement of Rome (whose own literary career expanded from copying letters to drafting his own), and others. But of all the documents in circulation, none were more prized than stories of Christ.

Capacity Problems

The earliest followers of Jesus swapped stories of his deeds and conversations. Wanting his memory to persist, they recited and recorded the principal events of his life and work: the archangel Gabriel's announcement of his birth, the commencement of his ministry, the Sermon on the Mount, the commissioning of the apostles, the feeding of the multitude, his confrontations with religious leaders, and his entry into Jerusalem, arrest, trial, execution, and ultimate vindication.

The first stories circulated orally but were soon written down. As Luke says at the start of his gospel, "Many have undertaken to set down an orderly account of the events that have been fulfilled among us, just as they were handed on to us by those who from the beginning were eyewitnesses and servants of the word."

For a time, both living and literary memories circulated simultaneously. Bishop Papias of Hierapolis in modern-day Turkey preferred the voice to the page. "Whenever someone arrived who had been a companion of one of the elders, I would carefully inquire after their words," he said. "I did not suppose that what came out of books would benefit me as much as that which came from a living and abiding voice."

It's easy to sympathize with Papias's ambivalence. Would you rather hear a story firsthand or take it from a book? But the question would soon be moot. All the eyewitnesses to Jesus's life would eventually die. Best to interview anyone remaining alive and record their stories. And that's exactly what people like Luke and even Papias himself did. The inquisitive bishop collected the statements and stories he discovered in a five-volume work called *Expositions of the Sayings of the Lord*, only the title and a few lines of which still exist.

Finally, four Gospels emerged as authoritative, those of Matthew, Mark, Luke, and John. These accounts were designed to encourage belief and settle differences about who Jesus was and what he was up to. As Luke says, continuing the quote above, "I too decided, after investigating everything carefully from the very first, to write an orderly account . . . so that you may know the truth concerning the things about which you have been instructed." And as John says in his gospel, "Now Jesus did many other signs in the presence of his disciples, which are not written in this book. But these are written so that you may come to believe that Jesus is the Messiah, the Son of God, and that through believing you may have life in his name."

Importantly, as John also said, "There are also many other things that Jesus did; if every one of them were written down, I suppose that the world itself could not contain the books that would be written." The sheer quantity of what did get written already posed a challenge. Scrolls can hold only so much before they become unwieldy. A gospel as long as Matthew or Luke would fill one lengthy scroll. But what if you wanted to have all four Gospels together?

When it came to their books, Christians possessed altogether different attitudes than their pagan neighbors. Despite Martial's support for the format, as we've seen, the ancients rarely used codices for literature. Instead, they were the tools of tradespeople, students, and authors working in drafts. Classical depictions of codices reveal humble uses: keeping records, storing reference work, or jotting notes.

T. C. Skeat, keeper of manuscripts at the British Museum, argued the codex's unique ability to accommodate larger collections of text, particularly the Gospels, drove the Christians' nearly wholesale use of that format, even for their most important writings. Other scholars have argued it was the collected letters of Paul that prompted the move from scroll to codex. A scribe required about eighty linear feet of papyrus split between several scrolls to reproduce Paul's letters; imagine Augustine lugging that around the garden. But that same scribe could fit all of his words in just one handy codex.

The additional space afforded by the codex could hold more than gospels and epistles. One third-century Egyptian codex contained copies of not only 1 and 2 Peter but also Psalms 33 and 34, Jude, the *Protoevangelium of James*, Melito's *Homily on Pascha*, and some apocryphal correspondence of Paul's: eleven different texts in all. Indeed, many benefits of swapping scrolls for codices—say, portability and easier reference—were probably afterthoughts. Instead, the sheer bookishness of early believers drove not only the change in format but also subsequent innovations in design and production to manage ever-larger quantities of work.

Take the use of quires. When pages are gathered and folded, they create a hinge along the fold: the spine. A single such gathering is called a *quire* (a *signature* in contemporary book production). The earliest codices were all single quires, dependent on just one hinge. But a single hinge limits the length of books. Too many pages, and the book becomes unwieldy. Worse, by overworking one hinge, the binding suffers, and the book falls apart. Christian bookmakers solved both problems by stacking many small quires and binding them together. Doing so created a wider spine with multiple hinges and could accommodate books of extraordinary length—such as all of Paul and one or more of the Gospels—in a single, handy volume.

The four evangelists. Collecting their words would help drive development of the codex.
SOURCE: YOGI BLACK, ALAMY.

"The impetus among early Christians for experimenting with these various techniques of codex construction was clearly to accommodate larger bodies of text into one physical book," says Larry Hurtado. "Their commitment to copying texts and to the codex book form drove their efforts to develop codex technology." But the ability to share large, collected texts was only part of the reason for the codex.

Another was in-group cultural reinforcement. At the end of the second letter to his disciple Timothy, Paul asked the young man to visit him. "Do your best to come to me soon," he said, adding, "When you come, bring the cloak that I left with Carpus at Troas, also the books, and above all the parchments." The word translated as "parchments" is *membranae*— the same term for notebooks, as we've seen. Early Christians used parchment notebooks to copy excerpts of Jewish scriptures useful for arguing their own interpretation of Christ's identity. Compiling these *testimonia* for regular reference was, as Gamble says, "in keeping with the ordinary use of the codex as a notebook."

The volumes Paul requested would have held his personal testimonia, notes, extracts, and copies of his letters, from which the final collection of his epistles likely came. Given that the primary use of the codex among pagans and Christians was so utilitarian, it's important that Christians opted to use the format for their most valuable writings. "Christian texts came to be inscribed in codices because they were practical books for everyday use: the handbooks, as it were, of the Christian community," says Gamble.

Burn It All Down

The Christians' pagan neighbors recognized their attachment to books. The satirist Lucian painted Christians as dupes for welcoming the huckster Peregrinus into their circle in Palestine. "He made them all look like children," he said, "for he was prophet, cult-leader, head of the synagogue, and everything, all by himself." Lucian credited Peregrinus's outsized influence to his mastery of texts. "He interpreted and explained some of their books and even composed many, and they revered him," he said. When authorities jailed Peregrinus in Syria, his Christian friends bribed the guards to let them visit. "Then elaborate meals were brought in, and sacred books of theirs were read aloud, and excellent Peregrinus . . . was called by them 'the new Socrates.'"

Peregrinus was jailed for his supposed faith. Christians opposed the ancient Roman pieties and refused to offer oblation to gods and emperors alike. Such observances theoretically secured the empire's order and prosperity, which meant Christian refusals endangered the peace of the community. The threat was initially slight because Christian populations were as well; in the year 100, Christians numbered fewer than ten thousand across the entire empire. There were perhaps barely twenty thousand when the prison doors clanked shut on Peregrinus in 132.

Official persecution in those years was only occasional, and legal action depended on public accusation or the direct intervention of the magistrate. There was no focus on Christians' reading habits—only on whether they would pay proper homage to the gods. But as the number of Christians grew throughout the second century, the scope widened.

When Lucian penned his takedown of Peregrinus in 165, there were as many as seventy thousand Christians, a number that had swelled to eighty thousand by the time pagan critic Celsus published his famous takedown, *True Doctrine*, in 170. Previous critics of Christianity based their appraisals on personal interactions. Celsus was among the first to engage with Christian literature directly, particularly the Gospels.

In 180, when Christians numbered more than one hundred thousand, a dozen from the North African village of Scillium were dragged before the Roman governor in Carthage. The governor, Saturninus, took interest in the portable bookcase present at the hearing. "What are the things in your chest?" he asked. "Books and epistles of Paul, a just man," answered one of the Christians.

It's unclear whether the Christians brought the books to defend their position or if the prosecution wanted to use the books against the defendants. Gamble notes it could have potentially gone either way. Not so during the Great Persecution under Emperors Diocletian and Galerius. By then there was only one possibility.

By the year 300, Christians amounted to one-tenth of the empire's total population. Any nonparticipation in pagan pieties now represented mass sedition. The situation was out of hand, and Christian literature was finally targeted for destruction. "An imperial decree was published everywhere, ordering the churches to be razed to the ground and the Scriptures destroyed by fire," says Eusebius.

Diocletian and Galerius personally oversaw the inaugural event in February 303 on the Feast of Terminalia. A church stood within eyeshot of the imperial palace at Nicomedia. Authorities stormed the building at dawn. "The gates having been forced open, they searched everywhere for an image of the Divinity," recorded the historian Lactantius. "The books of the Holy Scriptures were found, and they were committed to the flames." Religious paraphernalia of all sorts was rounded up and destroyed, especially books. The destruction swept across the empire, most brutally in the East. The only way to avoid death was to surrender your goods and resume Roman rites.

Three months into the Great Persecution, authorities in the North African city of Cirta showed up at one house where Christians met for worship. An official demanded the bishop hand over the church's books along with any other contraband. He surrendered items aplenty but no books. "The readers have the codices," explained the bishop. A search revealed barren bookshelves. But then another room was opened, and a subdeacon named Catullinus retrieved and surrendered a large book, probably a collection of the Gospels. "Why have you given only one codex?" asked the official. Even in the stark prose of the official record, you can sense he's losing patience: "Bring forth the scriptures which you have." But Catullinus and the others were empty-handed. "The readers have the codices," they said. "Show me the readers!" insisted the official. They refused and were arrested. But the authorities already had a list of the readers and began calling at their homes.

One after another, the readers surrendered what they had—four codices from the first, five from second, eight from the next, and so on, until thirty-seven books had been seized. Such seizures and surrenders were widespread and played into the later Donatist Controversy, in which, after the persecution ended, churches that refused to surrender their books and suffered the consequences refused fellowship with those that did fork over theirs.

On the Road, Again

Thankfully, the persecution did end. What's more, the formerly criminal books suddenly received a new and elevated status. Emperor Constantine came to power in 306. He not only decriminalized Christianity but also became a patron; many citizens of Constantinople followed suit.

By this point Eusebius had become something of a court intellectual. Constantine solicited his help to equip new churches with copies of the Christian scriptures. "I have thought it expedient," the emperor wrote to the bishop, "to instruct your Prudence to order fifty copies of the sacred Scriptures, the provision and use of which you know to be most needful for the instruction of the Church, to be written on prepared parchment in a legible manner, and in a convenient, portable form, by professional transcribers thoroughly practiced in their art."

Constantine said he would arrange for all the necessary materials, including Eusebius's use of two public carriages to deliver the Bibles across the same Roman roads that so many prior publishers of the good news had trod before. It was a rush job. "Take special care that they be completed with as little delay as possible," said the emperor. Eusebius went to work and fulfilled the order, delivering "magnificent and elaborately bound volumes." The lowly codex was now an official icon and symbol of—along with being the means to—cultural attainment and societal approbation.

Mary Beard, professor of classics at Cambridge University, attributes the ultimate success of Christianity to such traffic across this holy internet. "Christianity spread from its small-scale origins in Judaea largely because of the channels of communication across the Mediterranean world that the Roman Empire had opened up and because of the movement through those channels of people, goods, *books*, and ideas," she writes.

Henri-Jean Martin, professor at L'École Nationale des Chartes, says much the same: "As it spread through the Mediterranean basin with astonishing rapidity, Christianity maintained its unity and developed its doctrines only by recourse to writing."

Books not only fed this success but also embodied it. And ultimately the form of the book proved as significant as its contents. Larry Hurtado

Eusebius, historian, churchman, librarian. Engraving by André Thevet. *SOURCE:* CHRONICLE, ALAMY.

scoured the Leuven Database of Ancient Books and other ancient manuscript catalogs. In the second century, Christian books accounted for just 2 percent of all manuscripts. The tally rose to 10 percent in the third, a number that quadrupled in the fourth. And unlike classical books, most of the Christian books were codices. It's a roller-coaster trend line; scrolls plunge in the third and fourth centuries, while codices rocket upward in the fourth and fifth. In the year 100 almost all books were scrolls. By 500, nearly all were codices, and a hefty percentage of those were Christian.

Books are bodiless teachers. They live beyond the lives of their subjects and authors. And they extend the reach of those subjects and authors. "The book is a letter," wrote Florence Dupont, "and the letter is a traveler"—across not only space but also time. The ideas and arguments of one age become present and accessible to subsequent ages. We're suddenly back in the tablet archive of Suppiluliuma or sitting alongside Augustine in the garden.

Believers communed with God and his messengers through the books by and about them. As we'll see, that was especially true for the people who took on the work of producing those books.

Marginalia: Reimagining the Book

In the early fifth century, a wealthy Carthaginian man named Firmus heard part of Augustine's *The City of God* read aloud at a gathering. He'd already come across a few portions that were circulating and now wanted the rest. He wrote Augustine for a copy. The bishop was happy to oblige.

The City of God is a massive work, more than 350,000 words in English translation; by comparison, that's more than three and a half times the length of this book. Augustine divided the work into twenty-two books and wrote it over the span of many years. Since Firmus would need to have it all copied, how would he handle such a mass of words? The author had precise opinions.

"There are," Augustine wrote Firmus, "twenty-two books which are too bulky to bind into one volume. If you want two volumes they must be divided so that one volume has ten books and the other twelve." Augustine also allowed for another division of the books into five volumes; either way, he had a definitive sense of what his book was and how its arguments should be presented.

Many aspects of Augustine's letter to Firmus fit firmly in the late classical world to which it belongs. When he asks Firmus about distributing the book "to those who want to copy it" and limiting the number who could copy it simultaneously, he sounds very much like Cicero talking with Atticus about a literary project. Swap out the names, and the unknowing would never suspect hundreds of years stretched between one set of senders and recipients and the other. But there is also something new here.

Cicero and Atticus would have been dealing in scrolls. When Augustine mentions his "twenty-two books," he uses the term *quaterniones*, which refers to sheets of parchment or papyrus gathered, folded, and sewn into a quire. That is, we're clearly talking about codices, similarly obvious from Augustine's comment about binding the books into multiple volumes; not only does he actually use the term *codices*, but of course scrolls aren't bound.

The first codices, as we've seen, were single quires and could contain only so much text. But it wasn't long before bookmakers discovered how to stack and bind quires together and radically extend the capacity of a single volume—consider, for instance, Justinian's gargantuan sixth-century legal compilations.

Many of the first multi-quire codices simply compiled several freestanding works into one volume. But the development also made longer and more exhaustive works—such as *The City of God*—possible. That meant an author could work with a larger scope at a grander scale thanks to the codex and its evolution. Nor was this the only gain.

New benefits of the codex emerged as its use spread, especially the following:

1. *Convenience, including portability and economy:* Not only were codices easy to handle, but they could also be made relatively compact and were around 25 percent less expensive than scrolls.
2. *Sequentiality:* Codices could fix an order of component parts that had emergent meaning and interpretive power of its own. What's more, fixing the contents of books facilitates familiarity with texts and greater engagement.
3. *Ease of reference, including random access:* Thumbing through a codex is easier than rolling and unrolling a scroll. More important, bound pages allow a reader to jump in wherever they desire (or chance) and hold multiple places available for consultation.
4. *Quotability:* Ease of reference enabled scholars to take fuller advantage of the precise language writers employed to communicate their messages; readers could transcend the remembered gist of a passage and instead consult and reconvey exact quotations, enhancing critical engagement.
5. *Comprehensiveness:* The enlarged capacity of codices, coupled with the sequentiality already mentioned, made it possible to present several texts in a single volume, such as all of St. Paul's letters (as Augustine read in the garden) or several hundred years' worth of legislation (as Justinian compiled for his empire). But, perhaps more important, it also allowed authors like Augustine to make ambitious, exhaustive statements like *The City of God*, which, while tackling questions of history, ethics, philosophy, politics, and more, nonetheless can be presented as a unified message with the internal logic and integrity of a whole—a complete and compelling vision of the world.

The City of God represents something new in the world: the idea that a single book could deal with and present every fact or idea relevant to a certain aspect of life convincingly and authoritatively. What's more, the book could be accessible and, through its organization and production, even facilitate deeper access, its concepts available with just a casual browse or a vigorous riffle. Truth was only a few pages away.

What did Augustine bequeath to us moderns all these centuries later? "The idea," says his biographer James O'Donnell, "that wisdom, critically necessary wisdom, lies in the pages of a book. Antiquity invented the written word, but it was late antiquity that gave the written word its particular place of prominence." Find the right book, and you'd learn all you needed about heaven, hell, history, morals, war, taxes, cuisine, animal husbandry, math, grammar, rhetoric, ruling as a prince, sitting as a bishop. Today, O'Donnell says, that idea is on the wane, but it was "an immensely powerful shaping idea in all the time between."

All those times between are where we turn now, times that never would have taken the shape they did without the book, particularly the codex.

Chapter 7

WHAT THE MONKS DID

Where the Book Leads

The career of monastic pioneer Anthony the Great began with a book. Two, actually. As a young man, maybe just twenty, he walked to church in his village along the Nile. His parents had recently died and left him a large inheritance. What should he do with it?

Anthony couldn't shake the story from the Acts of the Apostles about the first Christians selling their property and sharing the proceeds. Pondering his next move as he walked to church, he arrived midway through a reading from the Gospel of Matthew. "If you wish to be perfect," he heard, "go sell everything you possess and give it to the poor and come, follow me." Now he knew. Persuaded by this pair of books, Anthony left, filled the pockets of the needy, and became a hermit.

Born in the middle of the third century, Anthony became a monk in 271. He lived long, dying in 356. A year later in 357, Athanasius, the bishop of Alexandria, published his story. *The Life of St. Anthony* was an instant success. Written in Greek, it was quickly translated into Latin and found a wide and eager audience. Based on what we know about book distribution at the time, this was something new. Augustine mentions hearing about it half a century later in his *Confessions*.

Anthony attracted followers almost immediately, and books were noticeably part of their lifestyle of self-renunciation and poverty. Athanasius said "holy bands of men" settled in the area around Anthony. Among other descriptors, he mentioned that these men "sang psalms" and "loved reading," though Anthony himself was likely less than fully literate (Athanasius described him as uneducated).

Another pioneer, the onetime soldier Pachomius, became a monk in 316. While Anthony preferred solitary monastic practice, with monks living largely on their own, Pachomius favored and eventually founded a communal form of monasticism. Initiates lived and worked together, cycling through daily set schedules of prayer and reading.

Literacy featured as a qualifier for membership. According to Pachomius's monastic rule, initiates first learned regular observances followed by select passages of scripture. What if the newcomer couldn't read? "If he is illiterate, he shall go at the first, third, and, sixth hours to someone who can teach and has been appointed for him," the rule states. "He shall stand before him and learn very studiously with all gratitude. . . . There shall be no one whatever in the monastery who does not learn to read and does not memorize something of the Scripture." Reluctant readers would be shown the door.

Naturally, a program geared toward reading housed books. Stewards watched over the library, which initially seemed small enough to keep in one alcove during the day and locked away at night. The library grew over time and included books both humble and exquisite.

As we saw earlier, reading had long marked a life of secular status and seriousness. Elite Roman culture, for instance, emerged from an intentionally literary education. "Every western aristocrat had to know Virgil by heart, and many other classical Latin authors," writes medieval historian Chris Wickham; "in the East it was Homer." The wealthy gained familiarity with their generation's great books in school and maintained it through intentional leisure—stretches of time away from the hustle and bustle of town and city, sequestered on country estates with family and friends, noodling on the verities.

Monks, however, represented something new—something different.

The Contemplative Life

Ancient literary culture was a product of privilege, though not exclusively so. In his book *On the Contemplative Life*, Philo of Alexandria described

The temptation of St. Anthony as imagined by Hieronymus Bosch. Look closely, and you'll see a book tucked behind the hermit's back. *SOURCE*: GL ARCHIVE, ALAMY.

a first-century community of religious recluses living in a sheltered community by the sea on what is now known as Lake Mariout in northern Egypt. Imagine a neighborhood planned by extreme introverts: They built homes in proximity, but not too close, and stayed indoors six days out of seven so they could be left alone to read.

"In each house there is a consecrated room which is called a sanctuary or closet [where] they are initiated into the mysteries of the sanctified life," explained Philo. "They read the Holy Scriptures and seek wisdom from their ancestral philosophy," he continued, along with "writings of men of old, the founders of their way of thinking, who left many memorials." The self-marginalized Dead Sea Scrolls community is similarly known for its literary activity—we know them principally by the books they left behind.

Economic constraints persisted, so when encountering examples of such literary devotion, we will often find upper-class devotees helping to foot the bill. But forget social status; the elevation and use of books became a significant marker of piety in the closing centuries of the classical period, especially for Christians—and especially for monks. "Always have a book in your hand and before your eyes," Jerome counseled one prospective monastic. Literacy was obligatory and spread with the monastic movement. Like the Pachomian program, the rule of St. Benedict presumed monks and nuns knew how to handle a book, and the rule of St. Caesarius of Arles mandated literacy for monks and nuns.

In his *Lausiac History*, Palladius described a nun named Silvania. "Being very learned and loving literature she turned night into day by perusing every writing of the ancient commentators, including 3,000,000 [lines] of Origen and 2,500,000 [lines] of Gregory, Stephen, Pierius, Basil, and other standard writers," he said. And she didn't just dip in and out: "She laboriously went through each book seven or eight times." This reading both lifted and liberated her. "Wherefore also," said Palladius, "she was enabled to be freed from knowledge falsely so called and to fly on wings, thanks to the grace of these books; elevated by kindly hopes she made herself a spiritual bird and journeyed to Christ."

Books took on symbolic and explanatory power for monks. For instance, lustful thoughts were compared in one saying of the Desert

Fathers to a book. If a monk remained unmoved by a lustful impulse, it was forgettable as an unpersuasive book. If entertained, however, the thought lodged in the mind like an unshakable conclusion. One Desert Father, Mark the Monk, likened self-reflection to reading a book. "The conscience is nature's book," he said. "The person who actively reads it experiences divine help."

Dorotheos of Gaza compared the spiritual life to learning to read as a boy—at first hard, then satisfying, finally all-consuming. "When I went to take up a book I was like someone going up to stroke a wild animal," he said of the difficulty. But perseverance paid off. "I became so engrossed in reading," he said, "that I did not know what I was eating or drinking, or how I slept, I was so enthused about my reading."

After his lessons, he would go home, bathe, and eat—his book always nearby.

> *I took whatever I found prepared for me, propped up a book beside me, and in a short time I was lost in it. For the siesta I had the same book as a companion by my chair, and if sleep overpowered me for a short time I was quickly on my feet again and at my reading. It was the same in the evening when I got back after lamplighting. I used to grasp my lamp and go on with my reading until midnight. So it was that I took no notice of, or pleasure in, anything except what I was reading.*

Dorotheos analogized this habit to the acquisition of virtue, which must become the monk's sole focus. But the model works precisely because Dorotheos's audience had an experience of reading close enough to his own that they could imagine being lost in their reading for days on end. Of course, the letter alone was dead; these words demanded action, just like those heard by Anthony all those years prior.

Program for Living

One monk read the scriptures day and night for two decades and then suddenly quit his home, sold his books, and left for the wilds. An elder monk stopped him and asked where he was going. "I have spent twenty years only hearing the words of the [sacred] books," he answered, "and now I finally want to make a part on putting into action what I have heard from the books." As Mark the Monk also said, "Words are not wiser than work."

Abba Serapion famously ate little and possessed next to nothing. He owned only the shirt on his back and a little Gospel book, probably Matthew's account. While walking in Alexandria, he once passed a poor man shivering on the ground. "How can I who am supposed to be a monk be wearing a smock while this pauper (or rather, Christ) is dying of cold?" he asked, referencing the Last Judgment depicted in Matthew 25 in which Christ identifies himself with the poor and judges his followers on their charity. Serapion took off his shirt and gave it to the poor man.

Sitting there naked with his Gospel, Serapion was soon approached by an official who knew him. "Who stripped you?" he asked, assuming the monk had been robbed. "This one stripped me," answered Serapion, holding out his Gospel book.

As Serapion left the area, he encountered a luckless debtor. What to do? Serapion sold his precious Gospel book so the man could pay his bills. When Serapion finally made it home, his disciple asked about the missing shirt and Gospel. Serapion answered by quoting from the book he'd just sold: "Well, naturally, he being the one who says to me every day, 'Sell all that you have and give to the poor,' I sold him [the Gospel] and gave the proceeds to him [the debtor], so that we shall enjoy greater freedom of speech with him at the Day of Judgement."

Following the teaching of the Gospel, Serapion identified both the pauper and the debtor as Christ himself and happily gave them his only and most valuable possessions. The verse quoted, significantly, was the same that had launched Anthony decades before.

Neglecting the holy books signaled lax spirituality, even apostasy. "The prophets made the books," said one elder; "our fathers came and practiced them. Those who [came] after those learned them by heart. Then there came this generation; they wrote them out then set them in the niches, unused."

A disciple of Abba Pambo stayed a couple weeks in Alexandria on business and slept at night in the entryway of the cathedral. He heard new and beautiful songs, composed for the church services there, and told Pambo about the experience when he returned home. The disciple wanted the local monks to sing the new songs as well.

Pambo disagreed. "The days will come when Christians will destroy the books of the holy gospels and of the holy apostles and of the divine prophets, smoothing away the holy Scriptures and writing troparia and pagan poems," Pambo warned. They will be, he said, "besotted with troparia and pagan poetry. This is why our fathers have said . . . not to write the lives and sayings of the holy fathers on parchment but on [papyrus], for the forthcoming generation is going to smooth away the lives of the holy fathers and write according to their own will."

Writing could be more easily scraped off the surface of parchment than papyrus, so the more durable material actually presented a threat to the longevity of the scriptures. Pambo feared that scribes who lacked parchment for their new poems would simply erase texts on hand, even the words of Jesus and Paul. But this worry, like most technophobic concerns, proved overblown. Over the next many centuries, monks were among the primary copyists and curators of the scriptures in the West— and books of all sorts, for that matter. And here the classical world finally flipped fully upside down.

So the Devil Finds You Busy

In the ancient world, slaves learned to read and write for the sake of their owners, not their own aims or enjoyment. Slaves were—returning to Rex Winsbury's description—the infrastructure of the book trade. The only status possessed by professional copyists was low. Monks, however, turned that situation to their advantage.

Reading and writing were arduous work, and the ancients denigrated labor. Very well, monks figured, they could deem and redeem the duty as a form of asceticism—just like fasting, almsgiving, and prayer. "By the middle of the fourth century, Christian sources are brimming with references to text transcription and dissemination," notes Kim Haines-Eitzen.

Why the uptick? The increase seems "a byproduct of the emergence of asceticism and monasticism, for we find that these movements effected a change in the notion of a scribe/copyist as low class: copying texts, and writing more generally, becomes an ascetic practice that raises one's religious stature."

Jerome, for instance, instructed his prospective monk to work "so that the devil may always find you busy." Among his recommended jobs? "Copy out manuscripts, so that your hand may earn you food and your soul be satisfied with reading."

Transcription's transition from the drudgery of slaves to an industry of monks happened as the classical period shaded into the early medieval period, hastened by the disruption of Roman schooling. Europe's intellectual center shifted from aristocrats to bishops and monastics. And while secular practitioners still copied manuscripts, monks now became the primary infrastructure of the trade.

Work abounded. "Christianity has no sacred tongue," writes historian Robert Louis Wilken, "but it cannot exist without books." Christian communities required tomes of all types: liturgical volumes for conducting services, collections of canons and rules, penitential manuals, saints' lives, biblical commentaries, and the biblical books themselves—not to mention the classical pagan works needed for rhetorical education and general learning and edification; Augustine identified those as essential, and his opinion was generally upheld. In a world before print, human hands scratched out every letter on every page.

Cassiodorus represents a transitional figure. A Roman patrician, he served under the Gothic King Theodoric and later retired to his family estate, the Vivarium. He built a monastery there famed for its literary activity. The work that "pleases me most," Cassiodorus told his monks, "is the work of the scribes," which he called "preach[ing] to men with the hand" and "fight[ing] with pen and ink against the unlawful snares of the devil." As monasteries spread and the need for books grew, *scriptoria*—manuscript-production facilities—sprouted up all across Europe.

Ezra the scribe as depicted in *Codex Amiatinus*. Many suppose the model was Cassiodorus at work in his library at the Vivarium. *SOURCE*: SMITH ARCHIVE, ALAMY.

Men predominated in the trade, but women prepared manuscripts as well. In 2014, researchers discovered lapis lazuli embedded in the teeth of a skeleton unearthed at a German women's monastery. A brilliant and costly blue ornamental stone mined in Afghanistan, lapis lazuli was pulverized to create ultramarine pigment for decorating expensive manuscripts. So how did this woman end up with bits of the stuff in her mouth?

"In adding detail to their illuminations, it is plausible to assume that artists would have occasionally licked their brushes to make a fine point, a practice that later artist manuals refer to explicitly," write the authors of an extensive paper on the discovery. "The repeated activity of inserting the tip of the brush into the mouth could explain the distribution pattern." The lapis lazuli particles suggest that this female scribe not only participated in manuscript production but also was highly skilled and thus entrusted with prized materials.

Nor was this nun alone. In his two-volume study published in 1896, *Books and Their Makers in the Middle Ages*, George Haven Putnam reviewed the data then available and said, "It is difficult to estimate too highly the extent of the services rendered by these feminine hands to learning and to history throughout the Middle Ages." Putnam recounted several examples of women scribes and convent scriptoria producing exquisite manuscripts. The examples have only multiplied in the years following, thanks to careful archival research and analysis of the manuscripts themselves.

Reduce, Reuse, Recycle

People commonly assume and say that medieval Christians were responsible for the destruction of the classical past, that books by pagan authors were targeted for elimination. And while you can find examples of intentional destruction, the truth is less scandalous. To survive through the centuries, books needed to be copied and recopied. That means scribes required fresh supplies of writing materials. Unfortunately, papyrus imports dwindled as Rome declined at the end of the classical era.

As the Middle Ages dawned, bookmakers responded by moving to parchment. Processed from animal hides, parchment naturally had limitations of its own; there are only so many calves and sheep running around at any given time. So for economic and practical reasons, parchment was scarce as well. And there were still more disruptions at play.

Around the same time the papyrus market evaporated, Christian preference for the codex dominated book production. The pagan past was almost entirely preserved on papyrus scrolls, but the future of the book for the next thousand years was parchment codices.

As the ancient books gradually fell out of use, monks failed to copy them, focusing their efforts and limited resources on copying books that mattered more to their particular communities. That usually meant more copies of St. Augustine and St. Cyprian than of Seneca and Ovid.

The attrition was slow but, given the incentives at work, also sure. And books that might have been preserved for a while could still end up lost because, as Abba Pambo had warned, parchment permitted reuse. The page of a book could be scraped or washed clean, allowing monks to make a fresh copy of another work.

"The obliteration of classical texts was not necessarily an act of hostility to the august literature of Rome," write scholars Andrew Pettegree and Arthur der Weduwen. To invoke a modern slogan: monks reduced, reused, recycled. Besides, as the authors say, "many more Christian than pagan texts were destroyed in this manner." Abba Pambo was right! But what's a monk to do when writing material is scarce and your spiritual practice requires copying? When you run out of Ovid to recycle, you turn to that extra copy of Cyprian—no slight intended toward either, just holy indifference.

Of course, Christian monks did preserve scores of pagan classical works. Whatever they found useful—mathematics, rhetoric, philosophy, and so on, even some Ovid—they kept and copied. That's why and how we still have those Latin classics today.

Code Upgrades

Monastics did more than copy manuscripts. They also improved the way books were written and read. Reading in the ancient world, as already observed, was cumbersome. When in the *Confessions* Augustine hears the command *tolle lege* (take and read), he could comply because he'd been trained to do so. Recipients of, say, Paul's letters did not have tidy paragraphs, sentences, punctuation, or even space between the words to help them; imagine the Epistle to the Romans, exceeding seven thousand words, full of layered clauses, complex phrasing, nuanced arguments, inelegant syntax, abundant digressions, and no place to even pause and breathe.

The voice told Augustine to *lege*, to read. This word shares a root with another more familiar to us, *elect*—to choose, to pick. Unlike Hebrew or Syriac, phonetic languages like Greek and Latin could get by without spaces or punctuation because readers could be trained to pick out the syllables by sounding them aloud. Readers didn't scan pages and identify words; they mouthed syllables. Hence Augustine's surprise, also in the *Confessions*, at seeing Bishop Ambrose reading quietly to himself.

Ancient readers usually read aloud. When Philip finds the Ethiopian eunuch on the road to Gaza, the eighth chapter of Acts says that he "*heard* him reading the prophet Isaiah." And when Cyril of Jerusalem directs the women in his church to "read silently," he has to qualify this instruction: "so that their lips speak, but others' ears catch not the sound." What astonished Augustine was Ambrose's ability to read without mouthing the syllables at all.

In Augustine's day and the centuries to follow, little in the text itself aided private reading. Instead, most people encountered texts read aloud in public settings. Augustine called the scripture the church's books; there are layers of significance to that phrase, but on the bare surface it refers to how people experienced the Bible. Most Christians heard the scripture at church, read aloud in services. Hence the promise of blessing in Revelation applying to both those who *read* and those who *hear* (chapter 1, verse 3). The assumption is not that people are reading the book in the privacy of their own homes but that they are gathered to hear a trained reader pick out the sounds so those who stand by and listen can understand the content of the book.

Appointed readers in each church or monastic community would carefully study the book to identify every syllable, a practice called the *praelectio* (prereading). The reader was responsible for picking out the syllables in a manuscript and deciphering the sense of the passage by sounding them out, arranging phrases and sentences and establishing the various meanings—all by uttering aloud and remembering the flow. He might mark up the manuscript with rudimentary punctuation to assist his memory during public recitations. Once fully deciphered and usually committed to memory, the message could be delivered to the congregation.

The work of a reader was a large responsibility. Scripta continua texts are inherently ambiguous. Since these are the words of God, clarity was essential. But it did not come through textual aids—like commas and periods and spaces between words. Rather, it came through rigorous training and received instruction. Teachers marked up manuscripts with crude punctuation to instruct their students regarding where to pause, what to stress, and when to stop. But those students wouldn't know where and how to employ these punctuation marks without the help of their teachers.

How big a problem was this? In Book 3 of *On Christian Doctrine*, Augustine devotes considerable space to the importance of readers properly pronouncing and punctuating the scripture. "When proper words make Scripture ambiguous, we must see in the first place that there is nothing wrong in our punctuation or pronunciation," he says, going on to give an example of how improperly punctuating the opening of John's Gospel leads to heresy.

Many without instruction went wildly astray. Irenaeus said the Gnostics misunderstood Paul because they tried reading the apostle on their own. Had they gone to church and listened to the readers' approved stresses and pauses, they would have received the proper sense. But since they approached the apostle without the benefit of the community and tradition in which his writings found their correct meaning, they flubbed it.

To ensure correct readings in these liturgical settings, leaving nothing to chance, Christian scribes broke with classical tradition early on and began innovating with punctuation. They did this by formalizing those marks teachers and other readers employed to decode books and quickly recall their *praelectio* work when standing and reciting the text in church.

Monk William of St. Martin of Tournai in bed, writing with six books open for reference. *SOURCE*: BIBLIOTHÈQUE MAZARINE, MS 753 F. 000IX. PORTAIL BIBLISSIMA.

Scribes working in monastic scriptoria began employing consistent word ordering, upper- and lower-case letters, chapter headings, paragraphs, ligatures, hyphens, periods, commas, and other forms of punctuation. Widespread adoption of these innovations took centuries, but they eventually became the norm.

No contribution proved more important than spaces between words. In Semitic languages, which lack written vowels, scribes placed spaces between words. Without those spaces, the page would be only a jumble of consonants. Irish (and later British) monks, familiar with Syriac manuscripts, took note. Following the Syriac practice—visible in the books they acquired from the East—they started inserting spaces between words when copying Greek and Latin manuscripts, placing stepping-stones in the river of letters flowing through their scripta continua texts. The earliest Greek manuscripts sporting spaces between words were of insular origin—centuries after those texts were originally written. Coupled with ecclesiastic and monastic punctuation innovations, spaces between words facilitated the kind of rapid, silent, individual reading we take for granted today.

The monks not only developed these lexical technologies but also employed them far and wide. A popular belief suggests Irish monks saved Western civilization by preserving and copying manuscripts. The Irish (and later Anglo-Saxons) really did save and spread classical, folk, and Christian literature. But their most astounding contribution, in conjunction with the Carolingian reformers who followed them, was developing textual and grammatical technologies that improved readability of those manuscripts—and every other book since—while encouraging literacy itself, something we'll soon explore.

Marginalia: Upping the Data Game

When Eusebius began to write his *History of the Church*, he was aware that he was up to something new. "I am the first to venture on such a project and to set out on what is indeed a lonely untrodden path," he stated up front. And the novelty in question? Unlike earlier historians, who tended to fabricate whatever data their story required—for instance, the lofty words of a general to his men—Eusebius composed his narrative from documentary evidence, texts within relatively easy grasp and ready consultation.

"I have," he explained, borrowing from Seneca and Macrobius, "picked out whatever seems relevant to the task I have undertaken, plucking like flowers in literary pastures the helpful contributions of earlier writers, to be embodied in the continuous narrative I have in mind." The churchman roamed a massive pasture: The episcopal library of Caesarea, from which he worked as bishop, contained thousands of volumes, both sacred and secular, stores that he augmented by scouring the episcopal library of Jerusalem, copying whatever he could.

Origen of Alexandria had planted the seeds of Eusebius's library a couple generations earlier. Raised a Christian and provided with an advanced education in both Christian and Greek philosophical thought, Origen proved book hungry his entire life: always buying, sometimes selling, and writing countless books of his own, especially pioneering examples of biblical commentary and theological speculation—working at one point with a large team of scribes, including, as Eusebius records, "girls trained for beautiful writing."

Origen settled in Caesarea, bringing his library with him and eventually folding it into the episcopal archive already there. A subsequent scholar, Eusebius's mentor Pamphilus, took the library as his special project with a focus on collecting, conserving, collating, copying, and cataloging. The end result was a research library envied and used by all around. Someone who required a copy of a work by Origen could find it at Caesarea. A scribe who wanted to correct his copy of a particular biblical book could compare it to copies available at Caesarea.

As Pamphilus's protégé and eventual bishop of the city, Eusebius inherited access to the collection and put it to rigorous use. His *History of the Church* is in many ways a history of books, lists, letters, and other documents. But his use of the library went even deeper, and innovative models emerged from its stores for sharing information that Eusebius refined and extended. What a library could do for data, an individual book could do as well.

Origen was the most adventurous and ambitious biblical scholar of his day. Recognizing the unique technological affordances of the codex, he pushed the innovation further than anyone before, most famously in his Hexapla edition of the Hebrew scriptures, "one of the single greatest monuments of Roman scholarship," according to book historians Anthony Grafton and Megan Williams.

The unique user interface of the page, as opposed to the scroll, permitted more discrete partitioning of information, allowing individual columns of text to serve individual purposes. Most codices of the time followed the simple format of one column per page—the dominant layout still used today. Yet pages will support multiple columns. A scribe could simply run a single stream of continuous text down the page in multiple columns, but Origen realized he could run multiple streams of text down the page in individual columns, permitting comparative analysis across the page—across both open pages, in fact.

Origen desired to see the Hebrew text side by side with a Greek transliteration of the Hebrew (Hebrew words in Greek characters), the Greek translation of the Hebrew known as the Septuagint, and three other Greek translations in circulation—those by Aquila, Symmachus, and Theodotion. The result? Six columns of text across two pages with each phrase broken into its own line, meant to be read both down and across, linearly and laterally—essentially the Bible as a spreadsheet, for line-by-line comparative analysis. It took forty volumes of roughly eight hundred pages apiece to produce, but the final result aided biblical research unlike anything else. Jerome used it not only to produce his biblical commentaries but also to help his new translation of the Bible into Latin (the Vulgate).

The original Hexapla resided in Eusebius's library at Caesarea, and its tabular presentation of data gave him an idea. What if you could display history visibly rather than narrate it linearly? The first part of Eusebius's *Chronicle* explains that the separate histories and calendars of individual nations don't stand alone; they overlap. Synchronisms occur when dates and events coincide. By identifying these synchronisms, a historian could construct a universal history, a timeline that encompassed multiple kingdoms.

Eusebius gathered chronological data on nineteen different kingdoms: Assyrian, Persian, Athenian, Jewish, Roman, and others. He then arranged it all by decades in a tabular timeline, so readers could quickly orient themselves and find what was happening across all these domains at any given time. The work proved so helpful that historians adapted and used it for more than a thousand years.

And he wasn't done innovating yet. Eusebius set his realization about tabular data to work on a problem that had vexed Christian scholars for centuries at that point: reconciling the divergent narratives of the four Gospels. By numbering the discrete sections of each Gospel, he could arrange tables of parallel passages, allowing a reader to quickly navigate between them—"the world's first hot links," as Grafton and Williams say. Eusebius's canon tables, as they were known, were used in biblical scholarship and illumination throughout the Middle Ages.

"No early creator of codices," write Grafton and Williams, "understood more vividly than Eusebius the possibilities that the new form of the book created for effective display of texts and information."

Chapter 8

GOD'S MOM AND BOOKS FOR ALL

The World's Most Famous Interruption

In the Annunciation, the archangel Gabriel interrupts the Virgin Mary to reveal that she will become the mother of the Messiah. But interrupts her doing what? St. Luke's Gospel account says only that Gabriel arrived and began talking; what Mary was up to goes unmentioned.

An early Christian writing known as the *Protoevangelium of James* provides Mary's backstory and offers a retelling of the Annunciation. It presents Mary spinning purple thread for a new veil in the Jerusalem temple. Gabriel's otherworldly voice sounds when Mary stops to get a drink of water. Unnerved, she returns to her seat and resumes her work, at which point Gabriel appears and relays his message.

Whether based directly on the *Protoevangelium* or on a prior shared tradition, many early visual representations of the Annunciation reveal Mary either holding thread or standing at a well with a pitcher. But later images in Europe, especially after the eleventh century, show something else: Gabriel interrupts Mary reading.

For a time, the two traditions coexisted. Some paintings depict a multitasking Mary, holding a book and her spindle. But eventually the book won out. By the late Middle Ages and the Renaissance, Mary is evidently more at ease in a library than beside a spinning wheel. As the tradition took hold, artists carried the theme beyond the Annunciation. New depictions appeared of younger Mary learning to read or older Mary teaching Christ to read, among others.

It's unlikely the historical Mary could read at all, but medieval Christians transformed her into an icon of literacy. Which leads to a question: When and how did the Mother of God become such a bookworm? It's worth considering because the image contains the story of medieval and early modern literacy.

Mary, the Model Reader

When Gabriel appears to tell Mary about becoming the mother of Jesus, she stops the angel to ask how. St. Ambrose, fourth-century bishop of Milan, said in his commentary on Luke that Mary knew Isaiah's prophecy that "a virgin [would] conceive and bear a son." Discovering she might be this selfsame virgin fired the young woman's curiosity, especially since Isaiah lacked detail. "Mary had read this passage," said Ambrose, "therefore she believed that the prophecy would come true, but she could not have read about how it would happen."

Notice Mary hadn't simply *heard* the prophecy; she had *read* it. Ambrose was one of the first people to refer to Mary's literacy. Nor was this the only time he did so. While praising Mary's virtues in his treatise *Concerning Virginity*, Ambrose said she was "devoted to reading." Even when she slept, "her spirit kept vigil . . . reviewing things she had read." Gabriel found Mary "alone in the most secluded room of the house, where she would not be distracted or disturbed."

Though alone, she wasn't by herself. "Indeed," Ambrose asked, "how could she have been alone when she enjoyed the company of so many books, so many archangels, so many prophets?" She communed with these and more through her prayers and reading.

Ambrose conjured this portrayal of Mary's studious solitude to encourage and instruct monastics, says University of Bergen professor Laura Saetveit Miles. A literate Mary modeled not only moral purity but also zeal for the scriptures and an interpretive mastery of the material. Miles highlights other writers who built on this portrayal in later centuries, including the Venerable Bede in the early eighth.

Like Ambrose, Bede assumed that Mary knew about the virgin birth from her reading but wondered how it would happen. "She was certain that what she was then hearing from the angel and what she had previously read in the sayings of the prophet necessarily had to be fulfilled," said Bede in his *Homilies on the Gospels*, "and so she inquired about the way in which it was to be accomplished. The prophet who predicted that this would be did not say how it could be done, reserving that instead for the angel to say."

So far these examples display Mary as a reader. But they also assume she had read the scriptures *before* the Annunciation. What's missing is any assertion that she was in the middle of reading when Gabriel paid his visit. That changed in the ninth century thanks to the widespread literacy campaign of a Frankish king with an inferiority complex and an anxious eye toward the End of Days.

The Emperor's New Books

Charlemagne's court at Aachen couldn't hide its rustic roots. Manners were minimal, learning low, refinement rare. The king's tour of the Italian peninsula drove home this point: The Franks were little better than barbarians by comparison with the Italians. More troubling, the clergy of his realm were poorly educated, prone to backward Latin and garbled recitation of the mass, laboring with error-riddled liturgical books.

Vexing under any circumstances, these deficiencies were doubly problematic because the king and his counselors believed they lived in the Last Days, an unruly time before the second coming of Christ. Threats to Christian Franks abounded. Heathen Vikings pressed in from the north, while Muslims pushed up from the south. And how else to explain false prophets popping up across the kingdom?

The last thing the Franks could afford was crumbling from within. Renewed armies of righteous soldiers could handle the military threat, and education would allow the faithful to refute the heretics.

Literacy had always been a priority of Christian settlements. Education, as we saw in the prior chapter, stood as a special concern for monasteries. Missionaries and priests taught letters along with prayers. In the sixth century, St. Ita of Ireland schooled local boys and girls in her monastery at Cell Ide, including a young St. Brendan the Navigator. In 669, the pope recruited Theodore of Tarsus and the African abbot Hadrian, freighted them with books, and sent them to Britain with a mission to evangelize and educate. Schools resulted, first in Canterbury, where Theodore became archbishop, and then at twin monasteries Wearmouth and Jarrow, which later combined. Wearmouth-Jarrow housed the industrious Bede. Meanwhile, the Irish brothers and sisters of St. Ita also busied themselves in Britain, setting up monasteries, scriptoria, and schools.

Unfortunately, these efforts succumbed to the depredations of Viking raiders, who destroyed monasteries along with the social improvement they offered. Fortunately for Charlemagne, he poached some of this insular talent before it was too late, including a star from the Wearmouth-Jarrow firmament: Alcuin of York.

Working with Alcuin and other advisers, Charlemagne issued two key reform directives. The first, *Letter on the Study of Literature*, in 784 briefly instructed clergy to take their own study and reading more seriously. "We exhort you," said the king, "to study earnestly in order that you may be able more easily and more correctly to penetrate the mysteries of the divine Scriptures. Since, moreover, images, tropes and similar figures are found in the sacred pages, no one doubts that each one in reading

these will understand the spiritual sense more quickly if previously he shall have been fully instructed in the mastery of letters." Only with greater intention and attention could they expect to master the material.

The second directive, tucked into a large ream of reform legislation titled *General Instruction*, came five years later. While covering many subjects, it drastically widened the scope for education: Charlemagne directed bishops and abbots to establish schools for the children in their areas and teach the psalms, stenography, chanting, reckoning liturgical seasons, and more advanced engagement with literature. The last point implied not only scripture but also pagan writers, without whom students would struggle to fully grasp the nuance and capacities of the Latin language.

Eager execution followed, fanning throughout the kingdom in local orders and commands. Evidence of success persists to the present. The majority of ancient Latin manuscripts available today were copied during the Carolingian age. Estimates of books copied in this period climb as high as fifty thousand. Roughly seven thousand of that total survive today, compared to around eighteen hundred for all the years prior. "Carolingian scribes were the unsung saviours of Western written culture," says historian Steven Roger Fischer. And new works were added to the mix. In the 250 years between 500 and 750, biblical commentators produced just 35 works; the tally jumped to 150 works in half that time following Charlemagne's campaign.

And what about Mary? With the tradition of Mary as a reader already in place, she soon became a poster girl for the reform movement.

Picture the Interruption

Around 860, a German monk at the abbey of Weissenburg wrote a fifteen-thousand-line vernacular gospel harmony in verse. Otfrid of Weissenburg is the first German poet known by name. He was the first to employ end-of-line rhyme. And in his *Evangelienbuch* he became the first writer to place a book in Mary's hand at the Annunciation.

Charlemagne receives a manuscript from Alcuin. *SOURCE:* HEMIS, ALAMY.

Describing Gabriel's approach, he said,

He found her praising God
With her psalter in her hands, singing through until the end.

Why the psalms and not, say, Isaiah, for which there was precedent in writers like Ambrose and Bede? Otfrid likely mined another source called the *Pseudo-Gospel of Matthew*, which expanded on the *Protoevangelium*; it mentions not only that Mary was learned in the law but also that she enjoyed singing psalms.

Along with this textual depiction came visual portrayals. Around the same time Otfrid penned his verses, roughly 115 miles to the west in Metz, an artist engraved an image of Mary with a book in hand on Gabriel's arrival. In the small ivory panel on what is known as the Brunswick Casket, Mary holds her thread in her left hand but also marks with her right thumb a page in an open book.

The world's most famous interruption: detail of Mary with an open book in Annunciation scene from ivory situla. *SOURCE*: METROPOLITAN MUSEUM OF ART.

"In the Brunswick Casket ivory carving," says Miles, "Mary clearly closes her hand around the top of the book in a gesture that evokes ownership, and also connects her more closely to the reading experience that was just interrupted: she keeps her page marked, perhaps her thumb on the line." A contemporary artist working in northern France carved the very same pose in a panel of an ivory pail—a situla—used to dispense holy water. The only difference is that the thread is gone. All the symbolism now resides in the book.

These visuals were picked up again a hundred years later by the English illuminator of the *Benedictional of St. Æthelwold*, which contains blessings for the bishop to offer for various feast days. The illustration of Mary and her book, patterned on Carolingian ivory carvings, comes before the benediction for the first Sunday in Advent. In this case, Mary's left hand clutches the thread, while her right rests on the page, her index finger holding her place. The image instantly calls back our picture of Augustine in the garden, marking his place in his book.

Mary was pictured as the great example of the contemplative life—the "faithful reader," as University of Ottawa professor David Lyle Jeffery put it. But if these words and images were largely meant for abbots, bishops, and other clerical professionals, as Miles indicates, how did Mary's bookishness spread?

An Upswell of Letters

While initially and primarily encouraging education for ecclesiastics, Charlemagne's efforts catalyzed literacy and learning across his realm and beyond. As the value of reading became more apparent, not only for its use but also for its prestige, lay people wanted in on the action as well. For some (say, those working in politics or law), it became a necessity.

Study of legal documents and charters, gift giving, entertainment, and spiritual pursuits reveals that literacy started among the nobility and trickled down. "The Carolingian laity, for a considerable way down the social scale, was a literate laity," writes Rosamond McKitterick, foremost expert on bookishness in the period. "The literate and the learned were respected; learning and literacy were aspired to, and became, in the

Mary holds her place: Annunciation scene from the *Benedictional of St. Æthelwold*.
SOURCE: BRITISH LIBRARY, ALAMY.

Carolingian world at least, a means of social advancement as well as of religious expression."

The Carolingian renewal, partially inspired by Anglo-Saxon monks like Alcuin, echoed back across the channel and helped inspire renewed educational efforts in England. In the ninth century, King Alfred the Great sponsored translation activity and advocated literacy among clerics and his nobles. Before Vikings had ravaged the countryside, "churches throughout the whole of England stood filled with treasures and books," recalled the king in a letter sent to accompany his personal translation of Pope St. Gregory the Great's *Book of Pastoral Care*. The trouble, as Alfred knew, was that knowledge of Latin had declined and learning decayed even before the Vikings' arrival. Few could read those books. Alfred conceived a plan for reform that would backfill the deficit.

It worked. And, as with the Carolingian reforms before, the effect reverberated throughout society. "By the twelfth century, when large numbers of texts and documents survive, there are references to the reading abilities of various kings, queens, noblemen, and noblewomen," says Exeter professor Nicholas Orme. "By the early thirteenth, it is manifest that towns too were centres of literacy. . . . By 1250, at the latest, the whole of the population was in contact with writings and literate people." This doesn't mean everyone could read. Only a minority could. But that minority was sufficiently diffused so that nearly everyone had exposure to them.

As the educational gains and practical benefits began to accrue, mass learning found more sponsors—including the pope. Abbeys and cathedrals had schools for local children, and schooling in the abbeys continued for monks who had taken their vows. Fees were required to attend, but, according to a study of English practice, at least some seats were reserved for families without means to pay.

These grammar schools provided basics in reading, writing, and mathematics. To go beyond the basics, universities began cropping up in the twelfth century, beginning with Bologna, Paris, and Oxford. While tensions sometimes existed between these early universities and church authorities, the church nonetheless supported their founding and expansion. "In an attempt to drive up the standards of monastic learning, in 1336 the Pope called on monasteries to send one in twenty of their monks for a higher education," writes historian of science Seb Falk. One ambitious abbey investigated by Falk pushed 15–20 percent of its members to the university. Others proved eager as well. "By 1500," says Falk, universities "had educated as many as a million students."

The content of a medieval education was communicated largely through books—studying, copying, disputing, and writing texts. Ascension through grammar school and university brought books into the orbit of students and graduates, beginning with Priscian's sixth-century go-to *Institutes of Grammar* and heading in whichever direction their studies took them. Aspiring astronomers would, for instance, begin studying Ptolemy's *Almagest*.

Entirely hand-copied with scarce materials, books were expensive. A lively market for these prized tools and resources arose in response. By the twelfth century, a "torrent of translations" washed across Europe, especially the works of Greek and Islamic scholars, spurring scientific inquiry as churchmen wrestled with the new insights and developed them further.

Female Literacy

Women participated in this upswell of letters. Queen St. Radegund founded the Abbey of the Holy Cross in Poitiers in the sixth century; her chosen monastic rule insisted all nuns learn to read and write. Radegund wrote poetry and was famous for her command of Greek and Latin church fathers. Tellingly, an eleventh-century depiction shows Radegund sitting with her writing tablets, evidence not so much of her advanced literary attainments (which were substantial) as of the Carolingian-inspired assumption of how a scholarly nun should appear: ready to write.

Evidence for female learning in this period is "maddeningly submerged from modern view," writes John Contreni of Purdue University; still, he says, what we do have "is far more impressive than that for both Roman classical and early Germanic societies." Memorization was basic to all—men, women, boys, girls—because everyone was expected to know the Lord's Prayer and the Nicene Creed by heart. Beyond that,

Queen St. Radegund with her writing tablets. *SOURCE*: POITIERS MUNICI-PAL LIBRARY. WIKIMEDIA COMMONS.

women monastics had to manage the same readerly efforts as their male counterparts. Not only did their monastic rules and profession mandate literacy, but, additionally, they were on their own: men were forbidden to spend time in convents, except to administer the sacraments.

Contreni points to signs of intellectual curiosity and significant projects from the period. Gisela and Rotrud, daughter and sister of Charlemagne, read Augustine's commentary on John for themselves and requested that Alcuin write another for them. Contreni also highlights cases of women teachers of both boys and girls. One famous example, the noblewoman Dhuoda, authored an instructional manual for her son early in the 840s full of spiritual, moral, and secular advice. Beyond that, women monastics typically copied their own books and books for others as well.

Hildegard of Bingen stands as one of the most celebrated and learned monastics of the High Middle Ages. She wrote three mystical-theological treatises, at least one of which won the approval of the pope and readers in every subsequent generation. Two scientific and medical treatises followed, along with two works of invented language, eighty or so songs, and a collection of nearly four hundred letters sent while she served as a Benedictine abbess to bishops, popes, secular rulers, monks, and fellow nuns.

Any of these works deserve recognition and exploration. But it's worth focusing on Hildegard's scientific and medical writings. Though primitive, her two books—*Physica* and *Causae et Curae*—nonetheless reflect immense learning and useful application. "The *Physica*, consisting of nine books listing almost a thousand plants and animals in German, is a study of botany, zoology, stones, metals and elements, describing their physical and medicinal properties," writes biographer Fiona Maddocks. "*Causae et Curae*, as its title indicates, examines the causes and cures of diseases . . . and offer[s] remedies, mainly using plants."

Hildegard was famous in her day for her curative methods, but she represents a wider movement toward medical advancement. Across Europe, "monasteries became centres of healing and medical expertise," writes Maddocks, "with their own elaborate herb gardens in which to grow their remedies."

This growing trend toward female literacy, spurred by both secular reforms and the rising number of convents and nuns, coupled with religious devotion to Mary, accelerated the image of Mary as reader. As these developments expanded, encountering Mary holding a book became commonplace. Artists began including the scene in altarpieces, stained-glass windows, and sculptures. By the end of the twelfth century, no one—priest, monk, nun, or layperson—remembered that the Theotokos had ever spun thread. She fingered texts, not textiles.

Read Like Mary, Be Like Mary

One reason for the change was the ability of the devout to identify with the Virgin by emulating her practice. As a result, depictions of the bookish Mary proliferated in books of hours. These private prayerbooks, favored first by upper- and later by middle-class lay readers, were popular enough to be characterized as "medieval best sellers" by the New York Metropolitan Museum of Art. Users were encouraged to see themselves kneeling in prayer just like the Virgin depicted.

"Any Book of Hours was liable to have a picture of the Annunciation in it," says religious historian Eamon Duffy. Here was the original scene of the Ave Maria, or Hail Mary, given on the same page of the prayer book that invoked it. "The female user of the book therefore no longer simply recites the Hail Mary," says Duffy, "she has climbed inside it, and has become part of the scene which her prayer evokes and commemorates."

This devotional identification with the Virgin was critical for the evolution of the Virgin as bookworm—and with greater literacy overall, especially among women. As religious reading grew in importance for late-medieval Christians, images of Mary reading inspired more such images until they came to dominate the iconography of the Annunciation, thus inspiring reading itself in a virtuous circle as the popularity grew.

Where did the Virgin Mary learn to read? The most common answer in the medieval West triggered another line of iconographic development: images of Mary's mother, St. Anne, teaching her daughter how to navigate a page. Some even show an elderly Anne with Mary teaching a toddler Christ to read; the three figures are collectively known as the St. Anne Trinity.

As with the earlier Annunciation images, these depictions rode a wave of rising literacy (especially as more women mastered their letters) and simultaneously encouraged its expansion. "Rather than simply mirroring the society of which it is a part, art functions to shape that society," explains Baruch College professor Pamela Sheingorn. "It is no accident or coincidence that the image of Anne teaching the Virgin Mary appeared when it did." As people in the late Middle Ages found literacy increasingly useful for reasons both religious and secular, Mary the bookworm emerged as the primary role model. To be like Mary was to read like Mary.

Monks and Carolingian reformers made it possible for average people to pick up a text and read it for themselves. But there were trade-offs. Making texts accessible meant that more people would access them, people with less consideration of the traditional understandings. The rapidly multiplying monastic libraries, writes Matthew Battles, represented "totalizing statements" of the tradition, the faith. But, he adds, "wherever writing"—and let's add reading here—"seems to achieve preeminence as a tool of the powerful, we find at that moment that it becomes possible to take it apart and turn it upon itself."

The innovations of the monks meant that an individual reader no longer had to endure painstaking instruction to learn how to read dense and difficult texts, "to hack their way through a thicket of letters, panting and uncertain, glancing back to make sure they didn't get lost," in Irene Vallejo's words. Now everyone could sit in a quiet room like Mary awaiting whatever muse might suit. Classical philosophy, folktales, even the words of God would soon be up for grabs. Charlemagne and the monks inadvertently democratized literature.

Jump back to the exchange between Theuth and Thamus: here we are again. Socrates's concern about untutored readers, never fully quieted, would sound loud again. Books of every kind would begin hobnobbing with readers of every kind, and those readers would interpret them however they chose. Eventually widespread personal reading would transform society.

Marginalia: One Tongue to Another

Despite his immense power and prestige, Abbāsid Caliph al-Ma'mūn stood awestruck. A ruddy-faced, handsome bald man with blue eyes sat before him. Al-Ma'mūn asked his identity.

"I am Aristotle," the man said.

"O wise man," said al-Ma'mūn, delighted with the revelation, "may I address a question to you?"

"Ask!"

"What is good?" asked the caliph.

"Whatever is good according to reason," said the philosopher.

"What else?"

"Whatever is good according to religious law."

"And what else?"

"Whatever society considers good."

"What else?"

"Nothing else."

This curious exchange reportedly happened during a dream, which is the only way it could have happened since Aristotle died around eleven hundred years before al-Ma'mūn was born. Ibn al-Nadīm, a tenth-century book dealer in Baghdad, tells the apocryphal story in his book *The Catalogue* to explain how so many books of ancient Greek philosophy and science ended up in the Islamic empire. "This dream was one of the most important causes," said al-Nadīm.

Al-Ma'mūn supposedly wrote to the Byzantine emperor about sending his scholars to libraries in Byzantine lands to find preeminent Greek books to bring home and translate. The emperor declined at first; al-Ma'mūn had only recently beaten him in battle. But he eventually relented, and al-Ma'mūn's team began scouting for worthy tomes to render into Arabic. The fact that Muslim scholars in Baghdad had already been translating Greek works with the help of Syrian Christians for several generations might lessen the veracity of the story but not its piquancy.

In the eighth, ninth, and tenth centuries, the works of Aristotle, Plato, Euclid, Ptolemy, Hippocrates, Galen, and many others (including Neoplatonic commentaries on authors such as Aristotle) eventually found their way into Arabic and traveled westward through Muslim North Africa and into Muslim Spain, especially the cities of Cordoba and Toledo. Along the way, Muslim scholars added to the bequest with commentaries on the classic works and advances of their own. The result was an enviable body of scholarship encompassing philosophy, mathematics, medicine, astronomy, and more.

European scholars soon realized what they were missing. The Christian kingdom of Castile took Toledo in 1085. Rumor of the literary treasures on offer quickly spread, and Latin scholars began migrating south to uncover whatever they could. One luminary among the influx was Gerard of Cremona, who arrived in Toledo around 1140 looking for a copy of Ptolemy's *Great Compilation*, more commonly known as the *Almagest*, thanks to its Arabic translation. He found it—and Aristotle and Euclid and Galen and more, not to mention scores of Arabic commentaries and original scholarship. Gerard learned Arabic for himself and translated more than seventy individual works into Latin. All these works joined the "torrent of translation" fueling the academic revival in Europe we covered earlier.

Not everyone liked it. On the one hand, some pushed back on the incorporation of pagan authorities into Christian education. Especially problematic? That ruddy-faced, handsome, bald man with blue eyes who dazzled al-Ma'mūn: Aristotle. On the other hand, some rejected the reliance on translations, insisting scholars should work only in the original languages. The English polymath Roger Bacon flew this flag.

"If I had control over the [translated] books of Aristotle," Bacon said, "I would have them all burned." His particular gripe was with University of Paris celebrity scholar Albert the Great, whose scholarship he found nearly as lamentable as his influence. "He has corrupted the study of philosophy," Bacon lamented. A significant part of the beef was that Albert worked from translations. Bacon warned that the translations then in vogue were riddled with errors, though he seems to have left no details about what those errors were.

But these are all questions of trade-offs, no? Go back to our idea grid. It's true that a poor translation might bump the expression and formulation of an idea down the x- and y-axes. But certainly moving from zero to something on either axis is a step in the right direction. Translation allows an idea formulated in one language to be recast in another and then expressed intelligibly, where before it would have only baffled and flummoxed. Besides, some of those errors might introduce what the painter Bob Ross was fond of calling "happy accidents." That might cause an original idea to suffer but birth something new and useful (consult the first- and second-century Christians who inherited the Jewish translation of Hebrew scriptures into Greek known as the Septuagint for a relevant opinion on that question).

Its drawbacks aside, the positives of translation have always outweighed the negatives, though there's also always been an allure to reading in the original languages, as our next chapter shows.

Chapter 9

HIT REFRESH

The Company of Ancients

After a change in Florentine leadership, the exiled civil servant Niccolò Machiavelli retired to his family estate outside Florence to await a change in fortune. Fending off boredom, he filled his days by managing the small farm, catching birds, chatting up the neighbors, and playing cards. Along with those humble pursuits, he read books.

"I go to a spring, and from there to one of my bird traps," he said in a letter, describing his typical day. "I have a book under my arm, either Dante, Petrarch, or one of the minor poets, Tibullus, Ovid, or the like. I read of their amorous passions and their loves, remember mine, and take pleasure for a while in these thoughts."

If you want a person to love reading as an adult, experts tell us, the best thing is to surround them with books as a child. When parents read and discuss books within view of their children, books transform into objects of interest; reading becomes a delectable pastime. Machiavelli's father, Bernardo, couldn't afford many books but purchased what he could: Aristotle, Cicero, and other Greek and Roman philosophers, along with books of rhetoric and Italian history. Books were costly at the time, and Bernardo expanded his humble collection by creative means. He once labored nine months compiling an index of Livy's *History of Rome* for the printer Niccolò di Lorenzo della Magna; in exchange for his trouble, he got to keep the book.

Machiavelli read his father's books. Biographer Maurizio Viroli says he was partial to ancient historians, including Thucydides, Plutarch,

Tacitus, and especially Livy. His interests also encompassed philosophy, comedy, and more. In his late twenties, Machiavelli copied a manuscript of Lucretius's *On the Nature of Things*, adding to the same codex a copy of Terence's popular comic play *The Eunuch*. Throughout his formative years, Machiavelli built an intellectual and emotional world saturated by the classical past; even his most beloved recent authors, such as Dante and Petrarch, reached back to antiquity for inspiration.

Throughout his forced retirement, Machiavelli found succor in this network of ancient ideas and associations. He hated being shunted aside. Back in Florence, he had associated with the influential; he undertook diplomatic missions and conversed with the powerful. Now he was nobody, doing nothing but catching birds and passing time with peasants. But his books! His books still welcomed him and affirmed his true identity and ambitions. You can see this in the letter describing his day:

> *When evening comes I return home and go into my study. At the door I take off my everyday clothes, covered with mud and dirt, and don garments of court and palace. Now garbed fittingly I step into the ancient courts of men of antiquity, where, received kindly, I partake of food that is for me alone and for which I was born, where I am not ashamed to converse with them and ask them the reasons for their actions. And they in their full humanity answer me. For four hours I feel no tedium and forget every anguish, not afraid of poverty, not terrified by death. I lose myself in them entirely.*

While Machiavelli was barred from the court in Florence, his books invited him to vicariously join the lofty and powerful once again. During these long, luxurious evenings, his experience of court life and personal reading fermented into something new: a little treatise for which Machiavelli would be forever remembered, *The Prince*. He said, "I have noted down how I have profited from their conversation"—that is, with his books—"and composed *De Principatibus*, a little study in which I probe as deeply as I can into deliberations on this subject."

He kept probing. While he despised his exile years, the period proved productive. He wrote not only the book for which he is known to the

Machiavelli with a book and his gloves. Painting by Santi di Tito. *SOURCE*: GL ARCHIVE, ALAMY.

present day but also his most important work of political philosophy, *Discourses on the First Ten Books of Titus Livy*. His dad's work on that index all those years before had paid dividends.

The *Discourses* represented an explicit attempt to use the past to inform—and reform—the present. For Machiavelli, voices echoing through the ages offered solutions to the conundrums of the day. And he was far from alone in this sentiment.

Discovering the Past

Charlemagne's educational initiatives inaugurated a centuries-long string of cultural renewal movements, culminating in the artistic and literary revival known as the Renaissance. Progress waxed and waned, but the trend toward heightened interest in literacy and learning proved irresistible and irreversible.

By the fourteenth century, rediscovery of classical texts characterized the trend. Take Livy. Few people read or cared about him during the medieval period. Monks had other worries than Roman wars and intrigue. Scribes in one French monastery scrubbed their rare copy clean and refilled the blank pages with St. Gregory the Great's *Moralia in Job*. Livy's complete history stretched to more than 142 books, so the likelihood of its being painstakingly hand copied and recopied through the centuries, when time and materials were scarce, was low to nil. Indeed, less than a quarter of the whole work survives. Without renewed interest in the fourteenth century, we'd probably possess far less.

As that renewed interest sparked and flared, scholars, poets, and collectors went hunting for copies. Often enough, they were church officials with access to monastic libraries: the same places that had rescued classical texts, including Livy, in prior days but now paid them less mind than Francis Petrarch and his contemporaries in the fourteenth century would have preferred.

Petrarch possessed a copy of the first ten books of Livy and was able to find, restore, and correct additional books of the collection, creating in his day one of the most complete and accurate manuscripts of Livy available. Even then exemplars and copies proved hard to come by; around a century after Petrarch, Antonio Beccadelli sold a farm to buy a copy.

Two interlocking drives fueled the manuscript hunt. First, curiosity and novelty combined with several early discoveries to make scholars thirsty for more. Aristotle, Euclid, Ptolemy, and others, sourced primarily from Muslim Spain, had already cut a wide path through European learning.

Second, as these rediscoveries and all-new discoveries mounted, especially by the fourteenth and fifteenth centuries, scholars and others began questioning the received picture of the world. The movement we know as Renaissance humanism was, say a trio of classicists, "motivated by a conscious break with the existing medieval cultural structure."

It was Petrarch who coined the term "Dark Ages" to characterize the prior centuries. Out with the old, in with the new—which was actually older. What else might be missing? What more waited to be found?

Petrarch woke many ancient manuscripts from hibernation, including several works by Cicero. Beyond his personal discoveries, he developed a network of manuscript hunters who scoured monastic and other ecclesial libraries from one end of Europe to the other in search of valuable documents. Sniffing out rare manuscripts grew into a hobby for many for the next few centuries.

Among the most famous manuscript hunters was papal secretary Poggio Bracciolini, a fifteenth-century contemporary of Beccadelli. Amusingly, Poggio once sold a copy of Livy to purchase a villa. "I want to know who," said Beccadelli in a letter to a friend, "did better, I or Poggio? He sold a Livy, beautifully written in his own hand, to buy a villa at Florence; I, to buy a Livy, have sold a farm." But Poggio was more apt to acquire than sell.

Poking around various monasteries, he uncovered several celebrated classical works, including orations by Cicero, Vitruvius's *On Architecture*, and Lucretius's *On the Nature of Things*, copied by Machiavelli all those years later. I was earlier able to quote from Quintilian's *Institutio Oratoria* because in 1416 Poggio found a copy moldering in a tower at the Abbey of St. Gall. Before that discovery, what few copies survived seemed fragmentary and incomplete. Petrarch never had more than a "torn and mangled" copy. Readers enthused over Poggio's find. "Our descendants will be able to live well and honorably," gushed the Venetian scholar Francesco

Barbaro in a letter to Poggio. Recirculating ancient wisdom for renewed application would grant improvement "to all mankind" and increase "the public good."

Inherent in all of Petrarch and Poggio's searches—and explicit in Barbaro's letter—is the reliance on books as tools for thinking new and different thoughts, something we covered earlier. Libraries are machines for thinking, and the humanists were itching to upgrade their software. They believed ideas found in classical manuscripts would better equip political leaders at all levels and ultimately benefit the whole of society; both Machiavelli's *The Prince* and his *Discourses* rest on this assumption. Unfortunately, the scope of this upgrade was inherently limited.

Greek to Me

Initially, most of the available classical manuscripts, even those rooted out of total obscurity, were in Latin. What's more, very few people in Western Europe could speak Greek. Petrarch tried learning, but, says historian Colin Wells, he "flunked." After receiving a copy of Homer's *Iliad* from a friend in Constantinople, Petrarch revealed his incapacity with the language while thanking his benefactor: "Your Homer is here with me; mute or rather in fact deaf am I, in front of him. But I am happy to gaze upon him and often hug him and, sighing, say, 'Great man, how I would love to hear you!'" Nearly as troublesome, because the ancient Latin texts the humanists were so eagerly rediscovering were usually built on Greek models, non–Greek speakers were doubly removed from a full appreciation of the legacy they sought to enliven.

A few ambitious Latins moved east to learn Greek, and some Greek counterparts moved west to teach. Schools developed around popular teachers, none more popular than Emmanuel Chrysoloras, who moved from Constantinople to Florence and on to Pavia and Milan, training students wherever he turned and writing a textbook for Greek instruction in the process.

Along with teaching Greek, Chrysoloras and his acolytes translated works by ancient pagans and Christians into Latin. Not only did Plato's *Republic* finally sound in Latin, but so did St. Basil the Great's *Address to Young Men on Greek Literature.* Because of earlier Latin translations, Aristotle had won fame among medieval thinkers. Now most of the humanists demoted him in favor of the newly available Plato.

Chrysoloras had company. Church councils in the 1430s brought Greek hierarchs, priests, and monks to western outposts such as Basel, Ferrara, and Florence to hash out doctrinal differences between the Latin and Greek churches on the outs with each other since the eleventh century. While Latin and Greek churchmen debated doctrine by day, some peeled off when possible to meet for more fruitful exchanges about books.

Ambrogio Traversari was able to poke through volumes belonging to the Greek scholar Bessarion. He identified mathematical works by Euclid and Ptolemy, as well as religious works such as Cyril of Alexandria's takedown of Emperor Julian the Apostate. Bessarion brought only part of his library with him, however. Traversari specifically mentions two books by the Greek geographer Strabo that Bessarion left behind. "How ill I took it that he had not brought the volumes along!" he exclaimed. "But I had to conceal the fact. I am led to hope, nevertheless, that they are to be brought."

They were. Bessarion continued to receive books with the intent of bringing his entire library to the West. "I assembled almost all the works of the wise men of Greece, especially those which were rare and difficult to find," he explained. "They must be preserved in a place that is both safe and accessible, for the general good of all readers, be they Greek or Latin." He eventually bequeathed them to St. Mark's in Venice, which he deemed "suitable and convenient" to Greek travelers.

Some of the Greek churchmen and their entourages stayed in the West, as Bessarion did. Working as teachers and translators, they helped expand Greek learning.

Not everyone involved in this movement saw eye to eye. In 1452, Poggio and George of Trebizond got in a fistfight over whether Plato was in fact superior to Aristotle. Poggio thought yes, George no. George went on to write a book against Plato: *A Comparison of the Philosophies of Aristotle and Plato.* Others joined the fight, trading paragraphs instead of punches. Bessarion defended Socrates's famous pupil with a book of his own: *Against Plato's Calumniators.*

BESSARIONI

Popes in this period became increasingly supportive of ancient learning. Nicholas V expanded the Vatican library with Greek classics and Latin translations of those works. Meanwhile, the Turks inadvertently encouraged the expansion of Greek by threatening and eventually conquering Constantinople in the middle of the fifteenth century. As Nicholas V busied himself filling Vatican shelves, books were busy fleeing Byzantium with their owners and looking for new homes. Greek-speaking scholars migrated westward, bringing large libraries of ancient Greek books, along with the ability to read them and teach others to do the same or translate them for those who could not.

Between 1397 and 1527, 165 different translators produced 784 translations of works by 154 Greek authors. During this period, pretty much everything known and available in Greek found its way into Latin. That included the foundational documents of the church itself, the Gospels and epistles of the New Testament. The urge to hit refresh eventually spread to the church, embodied most famously in two monks who both quit the cloister: Dutch humanist Desiderius Erasmus and German reformer Martin Luther.

Muddy Puddles and Rivers of Rolling Gold

Erasmus fully embraced the humanist calling. His *Adages* arranged pearls of classical wit and wisdom along with brief commentary. First published in 1500, the book stands as one of the major projects of his life, growing from just over eight hundred quotations in the first edition to more than four thousand in the final. We still use some of the expressions he included:

- "afraid of your shadow"
- "tip of the tongue"
- "teach an old dog new tricks"
- "leave no stone unturned"
- "break the ice"

Bessarion the bookish. Painting by Justus van Gent and Pedro Berruguete. *SOURCE*: PETER HORREE, ALAMY.

"The work would be a key means by which Renaissance Europe assimilated knowledge of the ancient world," says Michael Massing in *Fatal Discord*, his dual biography of Erasmus and Luther. The first edition featured sayings from Catullus, Cicero, Horace, Plato, Pliny, Terence, and others—about 80 percent Latin and 20 percent Greek.

Erasmus underrepresented Greeks because books in the language were still hard to come by, and he struggled with the tongue. "My readings in Greek all but crush my spirit," he confessed. Still, he considered knowledge of the language essential and knuckled down to acquire it. "Latin learning, rich as it is, is defective and incomplete without Greek," he said in a 1501 letter to Anthony of Bergen, abbot of the Abbey of St. Bertin, "for we have but a few small streams and muddy puddles, while they have the pure springs and rivers rolling gold." He considered it "utter madness" to theologize without Greek.

Whenever he had money, Erasmus purchased books in Greek before other necessities, including clothes. Imagine his delight when he discovered a cache of Greek volumes he could copy on the sly. He told Anthony about his find along with the tutor he hired to help him master the language, bragging that the man was "a real Greek, no, twice a Greek, always hungry, who charges an immoderate fee for his lessons."

The investment paid off. Erasmus became one of the most productive and popular humanist scholars of the age, proficient in Latin and Greek, and more widely published and read than most anyone else. During his life, more than a million copies of his books circulated through Europe, including his popular editions of the New Testament featuring his restored Greek text parallel with his fresh Latin translation.

Erasmus's entire enterprise was to bring the Latin church out of the muddy puddles and *ad fontes* back to the sources of rolling gold, to push through the overgrowth of medieval learning to the pure springs. Martin Luther agreed with Erasmus initially but soon broke in a more radical direction.

While studying law in his youth, Luther survived a violent thunderstorm during which he made a vow to become a monk. He joined the Augustinian order, where his overly scrupulous conscience both plagued him and tried the patience of his confessor. Like Augustine all those

Erasmus at work. Portrait by Quentin Metsys. *SOURCE*: GALLERIA NAZIONALE D'ARTE ANTICA, PALAZZO BARBERINI, ROME, ITALY. ARTEXPLORER, ALAMY.

centuries before, Luther found relief while flipping through the letters of St. Paul. Novel interpretations freed Luther from his tormenting conscience and provided a new theological trajectory.

The Roman church had novelties of its own, of course, including the sale of indulgences. Purchase of an indulgence promised to reduce time spent in purgatory, a way station in the afterlife where believers' sins were purged before their entry into Heaven. Luther took aim at the practice, drafting a lengthy series of theses—ninety-five in all—to dispute it. His move proved both catalytic and cataclysmic. Soon the papacy, the Holy Roman Empire, and various German principalities all erupted in response to this cantankerous monk.

As Luther continued to study and reinterpret, he deepened his commitment to his cause, using the relatively new printing press to revolutionary success. Working from the original Greek and Hebrew, Luther translated the Bible into the vernacular for greater use among average believers. He also published pamphlets and books on a wide range of subjects, outstripping Erasmus to become the most popular author in Europe.

Erasmus and Luther were just the crest of a wave. Many of the later reformers—for instance, John Calvin and Theodore Beza— were also trained humanists. In his *Tractionum Theologicarum*, Beza's third-most-referenced authority is Cicero, who racks up 197 citations. Vergil gets 110; Pliny, 72; Homer, 69; Plato, 56; Demosthenes, 49; Terence, 46; and Aristotle, 44, with a smattering of other Greek and Latin authors in the mix: Valla, Aristophanes, Sophocles, Galen, Ovid, Quintilian, Seneca, Varro, and more.

Beza, once away from his precious library for a few days, wrote an anxious poem about the separation, translated here by the Elizabethan playwright Thomas Heywood:

> *My* Cicero *and* Plinies *both,*
> *All haile to you; whom I was loath*
> *To leave one minut:* Cato, Columel,
> *My* Varro, Livy, *all are well.*
> *Hayle to my* Plautus, Terence *too,*

And Ovid *say, how dost thou doe?*
My Fabius, *my* Propertius,
And those not least belov'd of us,
Greeke Authors, exquisite all o're,
And whom I should have nam'd before,
Because of their Cothurnat straine,
And Homer *then, whom not in vaine,*
The people stil'd great: next I see
My Aristotle, *hayle to thee*
Plato, Tymæus, *and the rest*
Of you who cannot be exprest
In a phaleucik number; all,
Hayle to my Bookes in generall . . .

More than a series of name checks, Beza's list calls back to Machiavelli's bookish solace: the welcoming library, full of reassuring friends, voices from the past who helped make sense of the present and chart a course for the future. But let's not fool ourselves.

Transmission and Transformation

Whenever Machiavelli and Beza sat down among their books, their circle of counselors included more than Romans and Greeks. By necessity it also included the medieval monks, Muslim scholars, and Byzantine thinkers who preserved those books and also extended how they were used and understood.

Consider just the monks, whose influence we've already observed. While humanists failed to appreciate the significance of their efforts, monastic copyists radically altered the past by changing the technology that provided their access to it. With the monks' initiative and by their hands, scrolls had been transformed into codices. An entire way of reading and interacting with texts was swapped for another. The new model included books with spaces between words, increasingly standardized punctuation, improved scripts, and other upgrades and developments that rendered these books more accessible and useful.

In the thirteenth century, for instance, Robert Grosseteste of Oxford sought a means of retrieving his learnings from his vast range of reading. He came up with an extensive table—an index—of all the subjects he encountered in classical and patristic texts. That is to say, Grosseteste made himself an analog search engine. Dennis Duncan, lecturer in English at University College London, compared his index to "a Google on parchment that takes its subjects and explodes them across the whole of known literature."

Around the same time in Paris, Hugh of Saint-Cher came up with his own table. But instead of ranging across many books, his index focused on just one. "Hugh," says Duncan, "will be the first to produce a concordance to the Bible, to break the book down and rearrange it into an alphabetical index of its words." The final version identified "over 10,000 terms and listed them in alphabetical order." Duncan describes the concordance's random access to the Bible as "wormholes . . . through the scriptures." These and other medieval developments set the stage for Renaissance scholarship and would have been impossible without it.

Humanists prided themselves on going back to the sources, but the sources had been changed by the time they came along. Some had been lost, which was regrettable. But all that survived had been transformed by new and improved techniques and technologies. Humanists wanted to upgrade their software; they succeeded only because monks had already upgraded their hardware and made future developments feasible.

It wasn't long, for instance, before Robert Grosseteste's and Hugh of Saint-Cher's indexes morphed into the back-of-the-book technology that has smoothed the way for scholars ever since. To understand the idea machine, we must recognize the ways books and book technology developed alongside and with the content of books to create something new in the world. The hardware affects how we use the software, and vice versa. It's the interplay that explains our cultural evolution.

Nothing amplifies that point like the innovation of printing with movable type on paper. The arrival of print and paper not only facilitated the one-off successes of people like Erasmus and Luther but also absorbed all of the developments that had come before and triggered widespread transformations across European society.

Marginalia: Paper Pushers

Early Chinese books were engraved on strips of bamboo held together by cords. Bulky and heavy, such books were hardly convenient; knowledge was literally weighed and judged by the cartload. Silk also served for writing, but in the second century BCE craftspeople began experimenting with a new writing surface made by pulping fibrous plant material and allowing the fibers to float and interlace in water, bind together when drained, and then dry in sheets.

It took centuries to perfect the process, but paper was born and eventually became the preferred writing surface in China for state officials and religious adherents—almost everyone, really. Its appeal is easy to appreciate: Paper was relatively cheap and easy to make; it could be made from many substances, including hemp, bamboo, bark, even cast-off rags and old fishing nets; and it could be made in a wide range of qualities for a wide range of uses. Unsurprisingly, it also proved attractive beyond the borders of China.

The use of paper soon spread to central Asia along the Silk Road. Mark Kurlansky reports that the Hungarian-born British archaeologist Marc Aurel Stein found a cache of letters going back to the early fourth century. "I would rather be a dog's wife or a pig's wife than yours," an angry woman wrote to her husband when he left her without money in his absence. Importantly, she penned her gripe on paper. But the mass adoption and spread of paper outside China came from a much different source than peeved spouses.

To the west, Buddhists spread their scriptures on paper scrolls, not only into China but also throughout central Asia. Yet the greatest adoptee of the new technology was another faith born farther west. Muslim armies captured Chinese papermakers after a clash between their rival empires in 751. The captors put these craftsmen to work in Samarkand (in today's Uzbekistan) and learned the process for themselves, adapting it to the materials available in lands far beyond the source of its invention. They also adapted it to a new format picked up from their Christian neighbors and subjects: the codex.

The shift to paper not only coincided with but also helped facilitate an uptick in Islamic scholarship. As we've seen, papyrus and parchment presented very real supply challenges. In theory, paper also possessed upper-limit problems, but in practice, with a nearly endless supply of old rags to render for pulp, such limits posed no real issues. Quite the contrary: As the passion for translating and copying classical works took hold of Muslim scholars and quickly expanded into their generating original work of their own, paper offered a writing surface that could scale like nothing else.

A paper mill went up in Baghdad in 795. Within several hundred years, a preeminent theological school in the city had amassed a collection of some 140,000 books. Nothing in Europe could come close to that achievement. Meanwhile, in eleventh-century Cairo, one mosque library boasted a collection of 200,000 books, and just a century later a royal library in the same city housed eight times that number in a massive structure consisting of some forty rooms—1.6 million books. Nor was it just a few locales. Libraries worth bragging about were popping up all over Muslim-held territory, supported by heavy investment and endless reams of paper; to show just how deep the connection goes, we get the word *ream* from the Arabic word for bundle: *rizmah*.

Manuscripts at Santo Domingo de Silos in northern Spain reveal use of paper by European Christians in the tenth century, but, for reasons technical, practical, and in some cases prejudicial, most Europeans were slow adopters. "In 1231," writes Alexander Monro in his history *The Paper Trail*, "the Holy Roman Emperor, Frederick II, ruler of Italy, Germany, Burgundy, Naples and Sicily, banned the use of paper for all public notices and records in Naples, Sorrento and Amalfi, considering it less durable than vellum and parchment." He was pushing against the tide. According to Monro, government officials already favored paper over parchment, and by the time of the edict, use had been slowly spreading first by import and later by domestic manufacture.

The thirteenth and fourteenth centuries saw paper mills emerge all over Europe, first in Italy and then France and Germany. As intellectual and professional need for cheaper writing surfaces rose throughout Europe, demand for paper accelerated. Enterprising papermakers converted flour mills and took advantage of the abundant water to power their operations. The influx of papermakers drove down the price; soon paper was just a sixth the price of parchment, nearly 85 percent cheaper. Additionally, the upswell in supply meant European papermakers could return the favor to Arab customers, exporting their wares to Muslim countries.

More important, however, all those European papermakers were positioned to serve a new domestic market in the fifteenth century with the arrival of the printing press.

Chapter 10

TOO MANY BOOKS?

New Responses to New Realities

The thick-walled tower was originally built for defense atop a hill, but though wars persisted, Michel de Montaigne converted the third story of the squat, four-story structure into his library. With three large bay windows, it overlooked the rest of his house and property. "I am over the entrance," he reported in his *Essays*, "and see below me my garden, my farmyard, my courtyard, and into most of the parts of my house." As the tower was round, so was the library, built with specially curved shelves. "The only flat side," he said, was "the part needed for my table and chair; and curving round me it presents at a glance all my books."

As we saw in the previous chapter, Niccolò Machiavelli retreated into his library each evening while in exile. Montaigne, however, said he spent most of his time at home holed up with his books. Before converting the room into his library, "it was the most useless place in my house," but with the installation of five large shelves that could hold roughly a thousand volumes, Montaigne found himself thoroughly occupied inside. "In my library, I spend most of the days of my life and most of the hours of the day," he said. "There I leaf through now one book, now another, without order and without plan."

He described his method of working. "One moment I muse," he said, "another moment I write down or dictate, walking back and forth." Montaigne claimed he didn't read his books closely, but he kept some books handy for frequent reference and, based on his citations and the testimony of the *Essays* themselves, seemed to possess more than passing facility with many classical authors. Still, he disavowed any facility of memory and claimed he couldn't recall what he had written, let alone read. To compensate, he routinely scribbled marginalia in his books and even brief summaries of what he found useful in one or another book. "I have adopted the habit for some time now of adding at the end of each book (I mean of those that I intend to use only once) the time I finished reading it and the judgment I have derived of it as a whole," he said.

Montaigne's description of his working habits underscores a shift in intellectual life and engagement from the time of Machiavelli's exile three-quarters of a century earlier. To begin with, go back to the number of books on his shelves: roughly a thousand. Machiavelli had perhaps dozens at his disposal; his life bridges the transition period between hand-copied manuscript production and print. Machiavelli resorted to hand-copying books himself to add them to his personal stash.

Print, adapted from Chinese and Korean precedents with innovations pioneered by Johannes Gutenberg in the mid-fifteenth century, spread throughout Machiavelli's life, expanding the supply of books, and the growth of library holdings posed organizational challenges. There would be no need, for instance, to keep frequently used books close to hand in a collection of a few dozen titles—they were all handy in a set that small. But a thousand volumes presented a different challenge, even if the owner could see all his books at a glance, displayed in their cases and arrayed across his semicircular walls, as Montaigne could. The curator of such a collection had to impose order on the books, or the shelves would be chaotic.

What's more, the number of books he read and kept required Montaigne to work out a personal form of metadata to retain what he read and retrieve it without scanning or rereading the entire volume. Even in Montaigne's retirement from public life, he didn't have time for that; there were too many other books to read, thanks to his insatiable curiosity and the constant supply of new books pushed out by the presses. So Montaigne not only filled books with marginalia, as people always had done, but also added a note at the back on what he could find inside so he could pull it down and see at a glance whether it contained anything valuable.

Montaigne, essaying as usual. Sixteenth-century engraving. *SOURCE*: BIBLIOTHÈQUE
NATIONALE DE FRANCE. PHOTO12/ARCHIVES SNARK, ALAMY.

Scholars of all stripes were working out similar arrangements across Europe. As they navigated the post-Gutenberg world, they had to.

"A Phenomenon of Mass Mediatization"

In his celebrated debut novel *The Name of the Rose,* Umberto Eco describes a labyrinthine monastic library large enough to lose oneself in. Eco doesn't mention how many books fill its multiple rooms, but University of Sydney lecturer J. O. Ward worked out some straightforward calculations based on Eco's description of the layout and space. Given his numbers, the collection would amount to eighty-five thousand volumes, "the greatest in Christendom," as Eco said. Actually, far, far bigger. It's not uncommon for a modern American public library to contain so many volumes, but no medieval library in Europe possessed that number, not even the best endowed—nowhere close.

Going all the way back to the Pachomian project, many monastic libraries could fit their entire collections (probably just a couple dozen codices) in a niche in the wall or a single wooden box. As the centuries mounted, so did the stacks. Industrious monks added to the pile every year, but even then we're talking about relatively few volumes. In the sixth and seventh centuries, for instance, monks in the Latin West produced only around 120 books per year—in total. With such low output, libraries of massive size were unimaginable; most in Anglo-Saxon Britain, for instance, rarely exceeded sixty books.

Maybe we should expect small figures. This was, after all, before Charlemagne's educational reforms. But if we jump to the Continent a bit later in the ninth century, as the reformers beavered away and book production climbed, inventory evidence for monastic libraries still underwhelms. Saint-Riquier sported just 256 books; St. Gallen, 264; Murbach, 315; and Reichenau, 415. Even the best libraries were poorly endowed. Lorsch shelved 590 books, and Bobbio, one of the most preeminent of all, housed 666—perhaps an ominous number but still far short of what Montaigne, a single individual, not an entire institution, eventually amassed on his own.

One estimate pegs the typical library at three to four hundred volumes as late as the fourteenth century. "By the end of this period," says Ward, "exceptional libraries were inching over 1,000 titles." The Sorbonne, among the most exceptional, listed 2,066 volumes, though 300 of that total were missing and cataloged as lost; no wonder the 338 books on display were all chained to tables.

Can we think in totals? Numbers for the Latin West crunched by Utrecht University professors Eltjo Buringh and Jan Luiten van Zanden tally around 5.9 million books produced in all from the sixth through the fourteenth centuries. That's just seven thousand and change per year on average for the entirety of Western Europe. That number almost doubled in the fifteenth century as humanism took hold and professional copyists started churning out books too; almost five million books were added to the total in that century alone. But printing also arrived in the second half of that century and radically changed the math—along with everything else.

Between 1454 and 1500, more than twelve million books were printed, according to estimates by Buringh and van Zanden, and printers were just getting started. In the next century, the sixteenth, they pressed out more than two hundred million, and estimates can float as high as four hundred million. Between three and four hundred thousand individual works alone hit the market in the sixteenth century, a few of those landing on Machiavelli's shelves and many more on Montaigne's.

"By this art as much is produced in one day by one man even unskilled in letters, as it was barely possible to produce in a whole year by several men with the speediest quill," marveled the Swiss scientist Conrad Gessner. Writing in the mid-sixteenth century, while the eruption of print was still underway, he was awestruck by the prodigious output of the presses.

To stress the scope of the change Gessner described, examine the chart on page 162. In nine hundred years of manuscript production, from the sixth to the fifteenth centuries, European scribes produced about 10.9 million books in all. But between 1452 and 1600, printers produced some 212 million books (by conservative estimates), a 1,846 percent increase in a fraction of the time, less than 150 years. French historian Frédéric

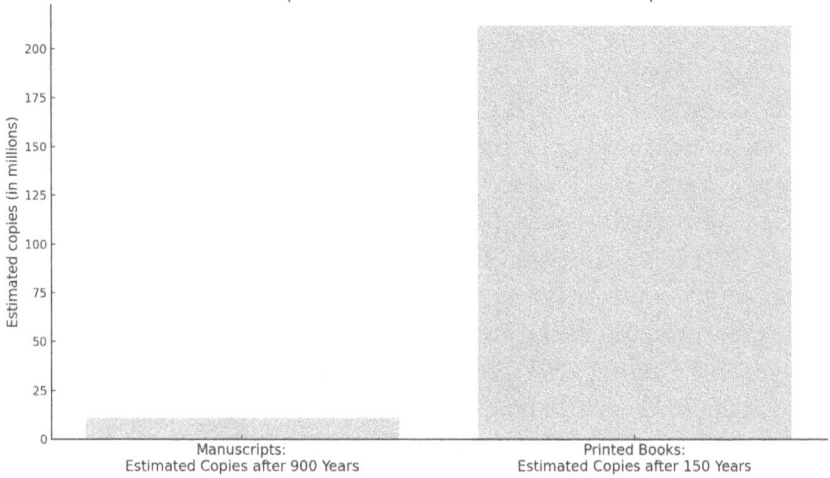

Manuscripts and Printed Books in Western Europe

Comparing manuscript production to printing. *SOURCE*: FROM ESTIMATES IN BURINGH AND VAN ZANDEN, "CHARTING THE 'RISE OF THE WEST'"; BURINGH, *MEDIEVAL MANUSCRIPT PRODUCTION IN THE LATIN WEST*; BARBIER, *GUTENBERG'S EUROPE*; AND FEBVRE AND MARTIN, *THE COMING OF THE BOOK*.

Barbier characterized this explosion of production as "a phenomenon of mass mediatization," one never before seen in human history. It had immediate ramifications.

The printing press presented the only way to fill an imaginary monastic library the size of Eco's, but printing would accelerate the trend begun by the humanists and push books well beyond the monastery. The only question was what to do with the surge: Some would seek to support it with new organizational structures; some would try to control the ideas it spread; others would destroy whatever they didn't approve. If knowledge is power, printing represented a massive expansion and redistribution of it.

Programming an Analog Search Engine

Back to Umberto Eco's impossible library. Another difficulty is finding anything amid such a gargantuan collection. Montaigne had to concoct homespun solutions for his thousand titles. What about something larger? Without an organizational structure and system for finding desired books or passages within books, "the user must look up and down all the shelves and books and read all the titles," as sixteenth-century librarian Juan Pérez said, implying the arduous futility of the work. Such a library would be, he said, "a dead library." Tellingly, Pérez made this statement about a collection not even a fifth the size of the library in *The Name of the Rose*.

Not only that, but Pérez's picture also assumed the frustrated library patron at least benefited from bookshelves. But shelves on which books were arranged upright with spines out were a recent invention pioneered by Pérez's own employer, Hernando Colón, bastard son of explorer Christopher Columbus. Eco set his tale in the fourteenth century, when librarians primarily stored their books in chests and arrayed them in stacks on benches or tables, making for tedious hunting. The situation was largely the same a century later.

Where the magic happens: inside an early print shop. Engraving by Jan Collaert I.
SOURCE: METROPOLITAN MUSEUM OF ART, HARRIS BRISBANE DICK FUND, 1934. PUBLIC DOMAIN.

Poet Antonio de Tomeis visited the Vatican library in 1477, during the papacy of Sixtus IV. According to his account, he saw "sixteen chests / Full of books." Those were topped with stacks of additional books, chained to ensure that they didn't wander off. If we assume one hundred books per chest and a few dozen atop them, we're talking about a library—*the* library at the nerve center of the Catholic Church—of just a couple thousand books. But the important detail is this: Shuffling through them all was grueling. When finished, Tomeis was "so exhausted and overwhelmed" that he had to "sit down, breathless."

By 1477, Tomeis was only a quarter century into the print revolution. Imagine the job with twice, thrice, or ten times the number of books. Imagine doing it with the eighty-five thousand supposedly in Eco's monastic library!

Large numbers of books have always presented difficulties for finding the helpful bits. We've seen how librarians in the Ancient Near East inscribed clay tablets with metadata such as a work's genre and incipit (the first few words of a text, usually to serve as a title), and in Alexandria librarians created a vast, information-dense, alphabetical catalog to navigate their holdings. Medieval scholars used and developed these and other techniques, employing both commonplace books—compilations of choice extracts from existing works—and indexes, as we saw with Robert Grosseteste and Hugh of Saint-Cher.

These ancient and medieval tools would be adopted and adapted for the age of print; recall that Machiavelli's father paid for his copy of Livy's *History of Rome* by compiling an index for the printer. No one developed the search apparatus of the library more spectacularly than did Juan Pérez's boss, Hernando Colón.

Enamored of the possibilities of print, Colón sought to amass a universal library. By the time of his death in 1539, he had collected more than fifteen thousand books in various languages—an astonishing number for the time—not to mention countless boxes of pamphlets, posters, and print ephemera that most regarded as worthless. Displaying the collection in his library in Seville, Spain, Colón aspired to construct the greatest idea machine ever built, featuring not only the best thinking of the human race but also its most diverse manifestation to date. But how to organize such a monstrosity?

Tired of rummaging through boxes of books or shuffling among the stacks atop benches, Colón championed the use of bookshelves with upright books, aligned and readable in rows, the way we organize books to this day. But his real advances lay in developing an analog network of hypertexts and hyperlinks that enabled users to navigate his massive collection of books, his *Libro de los Epítomes* (Book of Summaries) and *Libro de las Materias* (Book of Materials or Subjects), each stretching to multiple volumes.

With so many books to explore and limited hours to employ, readers would need a means of deciding whether this or that book warranted perusal. Since the title might be insufficient or misleading, the *Epítomes* provided one book that contained the general contents of many in summary form. It allowed a reader to get a sense of the subject matter and even, in many cases, gave details of the author's biography and style. What's more, these summaries were all cross-referenced with other catalogs Colón developed, essentially creating hyperlinks between volumes, a feature he developed even further in the *Materias*.

While a reader could beneficially browse the *Epítomes* at random, the real gains were for researchers or buyers who wanted distillations of books they were already curious about. But what if someone wondered about a certain topic and needed to know where to look for more? We would turn to Google or AI. Grosseteste and Hugh of Saint-Cher had similar challenges and created indexes that reached both across and within volumes to carve up knowledge into discrete conceptual packets that could be consulted as curiosity nudged.

Working within this tradition, Colón expanded it. He'd already added personalized indexes to certain books in his collection. Now he began creating an index that encompassed and cataloged the entire library—not just the broad subject matter of its books but all the individual topics covered within each, whether at length or in passing. So, as Colón's deputy Pérez explained, a person wanting to gin up a homily, a humanist oration, or a scientific treatise could work with whatever sources the library had to offer.

Hernando Colón's *Libro de los Epítomes,* containing his descriptions of roughly two thousand books. *SOURCE:* RITZAU, ALAMY.

Working with the *Materias*, the researcher could follow the trail of a subject through history, philosophy, theology, poetry, the Bible, whatever, freely across categorical distinctions such as author, book, and genre, a freedom that could increase the dimensionality and richness of the resultant research. "The *Libro de las Materias* created a network of connections between the words in the library," explain José María Pérez Fernández and Edward Wilson-Lee in *Hernando Colón's New World of Books.* Suddenly, the overwhelming ubiquity of books became not a burden but a benefit, as the index could help the reader find the choice bits he desired while encouraging serendipity along the way.

But though collectors and scholars like Colón reveled in the new challenges brought on by the flood of print, others fretted over what all those new books said, and indexes could be put to less scholarly and liberal ends than Colón ever imagined.

Silly, Useless, Idle, and Worse

The information glut occupied the minds of many. We've already encountered Conrad Gessner, the Swiss scientist who thrilled at the productivity of print. The godson of religious reformer Huldrych Zwingli and a friend of Theodore Beza, Gessner embraced the humanist project of recovering and preserving classical works. While still in his mid-twenties, he envisioned a Library of Universal Knowledge. Unlike Colón, he didn't try to assemble it himself, book by book. Instead, he hustled three years building an elaborate, annotated, alphabetical bibliography listing every known book in Latin, Greek, and Hebrew, including those long lost. He came away with a tome, the *Bibliotheca Universalis*, totaling thirteen hundred large pages and featuring some ten thousand works by three thousand authors.

Gessner likened his labor to wending through a maze or hiking a mountain. "I rejoice and give thanks to God immortal that I finally came out of this labyrinth," he said. "Now I look with joy on my labors and it pleases me to remember them: just like those who have come down a very high and steep mountain admire that they have returned when they are below again and congratulate themselves on overcoming a difficult path."

In some ways, Gessner's project mirrored Colón's, all the more so when three years later he produced his *Pandectarum*, a massive index to the *Bibliotheca Universalis*. Together the books would give librarians a catalog they could modify with call numbers for their own shelves, an acquisitions list when funds were available, and means for scholars to find needles in haystacks. As historian Ann Blair points out, Gessner saw a partnership between printers and librarians as a powerful tool for preserving his beloved classics.

But only those. Gessner openly fretted over the inflow of lowbrow books. "Although the typographical art seems to have been born for the conservation of books," he said in 1545, "most of the time nonetheless the silliness and useless writings of the men of our time are edited, to the neglect of the old and better ones."

With so many books available and more coming by the day, many scholars shared his general opinion. Several decades later, Thomas Bodley proposed rebuilding the library at Oxford. He kicked things off in 1602 with two thousand volumes; two decades later, it was ten times that size. Despite his fever-paced acquisition and the abundance that printing made possible, Bodley was choosy about what landed on his shelves. "Bodley envisioned that his library should be composed of serious books," say historians Andrew Pettegree and Arthur der Weduwen, "ordered according to the traditional hierarchy of university faculties: theology, jurisprudence, medicine and the higher arts." That meant, in Bodley's words, no "idle books and riffe raffes," such as, say, Shakespeare—or any other books in English. Bodley wanted only books in the classical and biblical languages.

Others fretted not over the language employed so much as the thoughts contained in all the books. Censorship in Europe goes back at least as far as 430 BCE when the Sophist philosopher Protagoras was hauled before a tribunal in Athens and convicted of impiety for suggesting in his book *On the Gods* that humans couldn't truly know whether his titular subjects existed. What exactly happened is unclear, but Protagoras either fled before the trial or was exiled. And his book? Supposedly burned. "If that indeed happened," says journalist and lawyer Eric Berkowitz, "it was likely the first book burning in Western history." It wouldn't be the last.

Sporadic attempts to suppress subversive ideas arose throughout the ancient and medieval world, though nothing systemic; there was neither need nor means. But the simultaneous rise of printing and the Protestant Reformation occasioned a different sort of response in which both sides of the conflict treated books and their contents like a battleground. What none of the participants appreciated at the time? The very forces driving the conflict made it impossible to contain.

Word Wars

Martin Luther's final break with Rome came in 1520 when he torched the papal bull that condemned him and his teaching. Supporters gathering for the protest also chucked Catholic pamphlets and books of canon law onto the pyre. This move set a pattern.

When anti-Catholic German peasants rose up in 1524, their first targets included monasteries; since monks were heavily involved in book production, the peasants singled out libraries for destruction. After one Cisterican monastery in Herrenalb was ravaged, no one could enter the ruins without finding shredded, dismembered books underfoot. Tallies were later made of the destruction at different locations; in some cases, thousands of books per monastery were destroyed, their shelves having been swollen by the recent arrival of print. Easy come, easy go.

While the mobs stripped the shelves in Germany, theft of monastic holdings in England was a formalized process, sanctioned by the government. After the Act of Supremacy in 1534, monasteries were dissolved and their property—including books—seized. The gentry and merchants bought up the assets and disposed of them however they saw fit. Collectors valued some of the books, but countless treasures ended up destroyed, used in some cases to rebind other books, clean boots, polish candlesticks, and serve even lesser ends; some found their way to lavatories, recycled one scratchy page at a time.

The ravages continued as the Reformation progressed. It became illegal in England to own Catholic books. Many books originally spared dismemberment and destructive reuse were later burned. Cambridge and Oxford saw their shelves picked clean of Catholic books during Edward VI's reign—and again of Protestant books during the reign of the Catholic monarch Mary Tudor. Hence the reason Thomas Bodley had occasion to exercise his snobbishness in restocking Oxford's library: The shelves he inherited had been stripped bare in all the fighting.

Catholic response to the Reformation could be every bit as brutal. In the Netherlands, for instance, printers and bookshops were raided, their wares tossed on pyres and sent up in smoke. Similar scenes played out across Europe, leading to—and furthered by—the various indexes of prohibited books that began proliferating in the mid-sixteenth century, empowering authorities to scour, purge, burn, and deface books by not only Protestants but also humanists, misbehaving or misbelieving Catholics, Jews, Muslims, and otherwise heterodox authors.

It started in the second quarter of the sixteenth century when various secular rulers, academic faculties, and church authorities began issuing lists of problematic authors and their writing. In 1526, for instance, England's Henry VIII published a list of forbidden books. There were only eighteen titles on the original list, three times as many books as Henry had wives. Arguably, he disliked the books even more than his unlucky spouses; after three years, he expanded the list to eighty-five titles.

France's King Francis I followed suit. In 1544, he authorized University of Paris theology professors to publish a list of heretical books, corrected and expanded every few years afterward. Book banners employed the same bibliographic techniques developed by Colón and Gessner but now to the opposite end, not to broaden but to restrict access and use. "Each edition's contents were arranged alphabetically, by last name of author, in two sections divided by language," says historian Robin Vose, "the first being entirely devoted to Latin works, followed by another in French (with some Italian as well)." Another section of the Paris index listed anonymous books alphabetically by title and language.

In 1551, the Spanish Inquisition used a similar list from the University of Leuven to create its index, which inquisitors employed in persecuting heretics. The Spanish list deviated from the Leuven original by including additional titles especially irksome to the compilers. Among the added works? Four by Erasmus.

This reactionary censoriousness played out across Europe with readily available authors finding themselves on the outs, including loyal Catholics like Erasmus. By the mid-1550s, as Vose explains, most Catholic regions possessed one or more such indexes, none of whose contents exactly matched. Lacking was a single, authoritative list and a centralized body to exercise regulatory authority. That came in 1557 following the accession of Pope Paul IV with the publication of the Vatican's official *Index of Prohibited Books*, which was soon revised and paired with an official oversight body, the Congregation of the Index, as part of the Tridentine reforms.

Who ended up on the naughty list? Protestant Reformers, naturally: Martin Luther, Martin Bucer, John Calvin, Huldrych Zwingli, and others. As mentioned, Erasmus made the list, as did Machiavelli. In time, so did Montaigne. In fact, looking over various editions of the *Index* is like reading a literary who's who. A sampling from this period and just beyond:

Francis Bacon	John Locke
Giovanni Casanova	John Milton
Daniel Defoe	Montesquieu
René Descartes	Blaise Pascal
Denis Diderot	François Rabelais
Gustave Flaubert	Jean-Jacques Rousseau
Thomas Hobbes	Baruch Spinoza
David Hume	Jonathan Swift
Immanuel Kant	Voltaire

Maddeningly, the *Index* provided no specific reasons for inclusion on the list. An author was merely listed or not. But as you can tell by these names, inclusion did little to squelch one's fame or appeal. If anything, it enhanced it. In 1627, Thomas Bodley's successor at Oxford, Thomas James, secured a copy of the *Index*, reprinted it, and encouraged librarians to use it as a theological purchase guide.

If discovered, listed books might be burned or, if only marginally offensive, redacted or otherwise purged of problematic passages by scribbling them out, painting them over, or tearing out pages. One section of a 1541 copy of Erasmus's *Adages* "was treated with particular disdain," writes Owen Jarus of *Live Science*, "having pages ripped out, sections inked out and two of the pages actually glued together."

But the censors were hardly consistent. When, for instance, they "corrected" Giovanni Boccaccio's humanist masterpiece *The Decameron*, widely loved for its language, craft, and inventiveness, censors left its obscenity untouched and instead edited the bits that made clergy look bad. They could be brutal: Venetian book smugglers Pietro Longho and Girolamo Donzellini were both seized by authorities, trussed up, and drowned at sea.

The trouble? Whether doctoring individual copies, bowdlerizing new editions for print, or punishing scofflaws, the censors couldn't keep pace with the printers. Recall the numbers at the beginning of the chapter: Printers produced between three and four hundred thousand individual

Jehan Georges Vibert's painting *The Committee on Moral Books*. *SOURCE:* WIKIMEDIA COMMONS.

works in the sixteenth century alone, tallying between two and four hundred million copies. The idea that censors could read everything produced, let alone scribble out individual lines, is laughable.

The *Index of Prohibited Books* was a medieval answer to a modern problem and failed to meet the challenge. Instead, the explosion of books and literacy occasioned by the printing press was about to reshape the world, as evidenced by the names of many of those prohibited authors.

Changes on the Horizon

Whereas books could be read and accessed by only a select few in the ancient world, innovations in the classical and medieval period—everything from the codex to spaces between words—improved the use and accessibility of books. As printing put books in the hands of the masses, they would change the trajectory of science, politics, civil rights, personal psychology, and more.

Marginalia: Readers Unleashed

At the close of the fifteenth century, Ottoman authorities banned printing for all Muslims living within the empire, underscoring the decision with subsequent decrees. "In 1485 Sultan Bayezid II issued an edict prohibiting the printing of books in Arabic script," writes Kenyan scholar Calestous Juma. "In 1508 the Shaykh al-Islam of the Ottoman Empire—the highest-ranking Islamic scholar—issued a fatwa stating that printing using movable type was permitted for non-Muslim communities but not for Muslims." Bayezid's successor, Selim I, doubled down; for Muslims, fooling around with a printing press could result in the death penalty.

What's remarkable about the Muslim rejection of printing? Like its cousin, Christianity, Islam produced a book-centered culture—not only the Quran but also, as we've already seen, untold works of commentary, philosophy, science, and other literature. What's more, Muslims encountered Chinese printing techniques long before Gutenberg set up shop in Strasbourg and Mainz, Germany. One might assume an eager adoption of the new technology. But no.

Juma links the rejection of printing to worry about its potential to undermine Islam's oral literary culture (which hand-copied books were seen to support rather than subvert) and fears that printing would erode traditional forms of intellectual authority, not only religious but also secular. "The great increase in books," Juma quotes one writer as saying, "made it possible for ignoramuses to infiltrate the ranks of the qualified intellectuals."

These concerns echo Socrates's complaint eighteen centuries before, and the European experience validates the worry.

As we've seen, Christians traded their scrolls for codices early on. While the new faith waxed and the old structures waned, the novel format took hold wherever learned people collected and used books, especially since the majority of those who did so were either monks or otherwise connected to the church—men such as Origen, Eusebius, Jerome, and Augustine. Embracing the innovation of the codex, the monks and churchmen innovated further, experimenting with page design and layout, script design, punctuation, spaces between words, search tools such as indexes, and more. All of these developments enhanced the usability of books and prepared the way for heightened levels of literacy and literary engagement.

The largely bookless religions of ancient Rome cared little for orthodoxy—that is, a faith with consequential propositions to which adherents must assent. It was enough that people performed the required rites. But religions of the book—Judaism, Christianity, and Islam—tend to demand some degree of alignment with ideas either derived from or reflected in their divine texts—namely, the Hebrew Torah, the Old and New Testaments, and the Quran.

A book provides an anchor to moor a community to a certain understanding of the world. But it also provides a means to unsettle that mooring and send the community adrift in different directions. Islamic authorities recognized the potential for the printing press to destabilize their community and banned it. Christians welcomed the printing press and soon found their entire civilization ruptured.

While one of the first uses of the printing press in Europe was to produce certificates of indulgence—a practice that served the Roman Catholic Church—the emerging Protestant movement soon employed the press to attack indulgences and eventually the very foundations of the Catholic Church. No one embodies this shift better than the reformer Martin Luther, whom we met earlier. But a singular focus on Luther obscures a dynamic from which he benefited but that ultimately caused the Reformation he began to spin out of his control.

Socrates feared that broad access to books would democratize not just knowledge but also, and more important, interpretation. Where the head of a reading community can direct or control the interpretation of the community's book(s), he can ensure that a certain understanding holds sway. Want to know what to think about a, b, or c? Consult the head of the philosophical school, the rabbi, the bishop, or the imam. But in fulfillment of Socrates's fear, enacting the same worry intuited by Muslim authorities, the printing press encouraged competing interpretations.

During the early days of the Reformation, Luther led the charge. His interpretations of the New Testament flatly contradicted what the bishops of the Catholic Church claimed. And he didn't simply say so from the pulpit. He also published an avalanche of pages to the same effect, utterly dependent on all the innovations to the shape, form, and production of the book we've studied so far, not least paper and printing, which permitted scale unlike any previous literary innovations.

It's too simplistic to say these technologies produced the Reformation, but they were necessary (if not sufficient) causes, and Luther wasn't the only person with access to reams of paper or a press. Once the bishops were rebuffed, other reformers could read the same Bible that Luther could and come to different opinions about its meaning—and they did. While the vast majority of Europeans were illiterate, by the sixteenth century more could read than ever before. They could also write.

As vernacular Bibles mushroomed up across Europe, so did the number of interpretations popularized by self-appointed authorities who could now compete in the emerging marketplace of ideas. Christendom shattered like a glass hurled into a fireplace.

This newly unleashed power of individual interpreters represented either blessing or curse, depending on which team a person rooted for. The Catholic Church soon regretted its early laissez-faire attitude about the press and instituted its failed *Index of Prohibited Books*. Should it, could it have moved earlier, as the Muslims did? Would it have mattered?

Islamic authorities eventually liberalized their stance on printing. By then, however, much of the Muslim world's head start in science, medicine, philosophy, and other areas had been squandered and Europe had pulled far ahead, powered in large part by individuals substantially free to pursue their own intellectual projects, fueled by books.

APPLICATIONS

Information technologies . . . modify how we act in the world; they also profoundly affect how we understand the world, how we relate to it, how we see ourselves, how we interact with each other, and how our hopes for a better future are shaped.

—Luciano Floridi

All machines have their friction.

—Henry David Thoreau

Chapter 11

READING THE BOOK OF NATURE

Notes on a Theory

When Cambridge University librarian Jessica Gardner peeked inside the garish pink gift bag left outside her office in the spring of 2022, she found something for which the library had been hunting for more than two decades. Anxiety welled as she opened the package. "I was shaking," she admitted. Within the bag was a brown paper envelope.

"Librarian," it said, "Happy Easter," signed X.

Inside the envelope? A small blue cardboard box enclosing two books wrapped in clear plastic. Gardner suspected what they were but couldn't afford to get her hopes up. "I was . . . cautious because until we could unwrap them, you can't be 100% sure," she said. Police inspected the package and finally gave the go-ahead to open the books five days later.

It's no wonder the drop was signed X. If you'd stolen the notebooks of nineteenth-century naturalist Charles Darwin, or had any connection to the person who did, you'd likely want to remain anonymous in hopes of escaping prosecution. Worth millions of dollars, the little leather-bound books contain the scientist's notes on what he called transmutation. "They show," said Gardner, "the evolution of Darwin's theory of evolution as he's thinking on the page." One of the pages in the returned notebook featured Darwin's now famous tree-of-life sketch in which he visually represented descent from a common ancestor.

Darwin's theory emerged from years of thinking on the question of origins, and his notes don't simply record his thinking. Per the insights covered earlier, they *are* his thinking, part of the actual process of nood-ling it all out—"thinking on the page," as Gardner put it. "He wrote the notebook entries in telegraphic style," said biographer David Quammen, "without much concern for punctuation or grammar. There were inser-tions, cross-outs, abbreviations, bad spellings. . . . He was brainstorming." The incubation period proved lengthy; it took more than two decades for Darwin's thinking to coalesce as *On the Origin of Species*.

Isaac Newton worked much the same way—though faster and also, curiously, even slower. He suspected gravity's tug explained more than why, say, an apple might fall from a tree; he was sure it explained the movement of all the celestial bodies overhead. But how to understand and describe it?

Newton meandered head down in his garden, turning on the ques-tion as he turned through the foliage. Ideas came in sudden jolts. His assistant reported seeing him shuffle along and then jerk himself upright and race back to his desk to write down whatever idea had flashed to mind before it escaped him. He stood at his desk and scribbled away, trying to stretch and knead every thought. According to his biogra-pher James Gleick, "Thousands of sheets of manuscript lay all around." This scene and others like it repeated countless times over the eighteen months Newton worked on the *Principia Mathematica*, a lengthy book that revolutionized human understanding of the invisible forces at work in the natural world.

Having teased these thoughts from the turning of his mind, Newton then evaluated and improved them on paper. In 1687, he published the initial result of his labors. He released a second edition of the *Principia* in 1713 after reshaping what he deemed faulty or infelicitous passages. Like Darwin's notebooks, Newton's pages feature scribbles and cross-outs, revisions of revisions. And he kept on revising, hunting for the ideal formulation for his ideas. In 1726—almost four decades after the first edition—Newton issued a third edition with further refinements and updates.

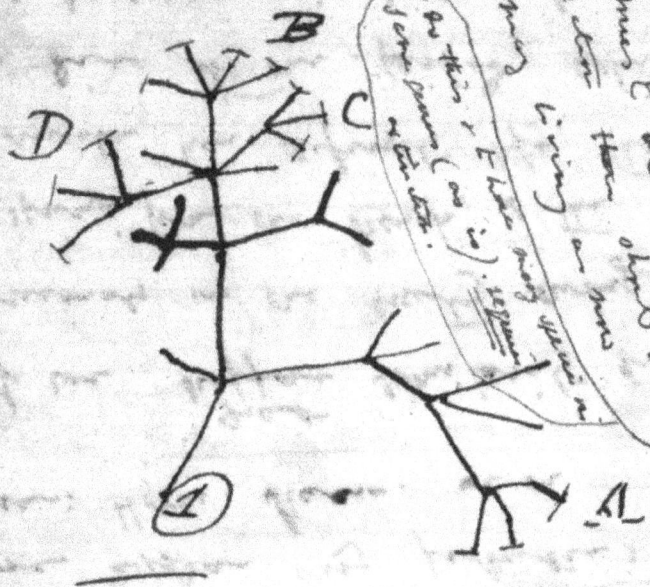

Darwin's tree-of-life sketch from his notebooks, taking his hunch one step closer toward a specific formulation of a theory. *SOURCE:* WIKIMEDIA COMMONS.

Newton's preparations for a second edition of his *Principia Mathematica*. Yes, those are his revisions. *SOURCE*: BILLTHOM, CREATIVE COMMONS.

Two Ingredients

Humans have been explaining the world to each other since the start. But science goes beyond mere explanation. Ancient cultures tended to attribute the world and its peculiarities to the interventions of one sort of divine being or another. While not eschewing the possibility of divine presence or even activity, science, as we recognize it today, relies solely on naturalistic explanations and regards spiritual forces as outside its purview. But methodological naturalism alone isn't enough for science to emerge.

In the sixth century BCE, Anaximander of Miletus became, so far as we know, the first person to describe the earth as an object—in his case, a cylinder rather than a sphere—suspended in space. What's more, he devised entirely natural explanations for meteorological phenomena, the appearance of the sun and moon, and more, even the emergence of life. But as historian Colin Wells says of his notions of a free-floating earth, "The important thing isn't really that Anaximander had the radically new idea." After all, others may well have come up with similar schemes. "The important thing," he says, "is that we have it." Anaximander wrote a book. Though now all but entirely lost to us, Anaximander's *On Nature* provides a glimpse of the first science.

Science requires two primary ingredients: naturalistic explanations of phenomena and written communication. Anaximander was lucky enough to live when his ideas could be captured and refined in writing and then passed along to others who could engage with those ideas, including those who would come long, long after. Anaximander underscores the point in an additional way. He was likely the first philosopher to write in prose. Before him, philosophers wrote in poetry, which might have been more beautiful or uplifting than prose. But prose offered the benefit of greater precision of expression, necessary for the finer points of theorizing and description.

The history of science is the history of books in various states of dress (loose notes, first drafts, second editions) and their readers (supporters, critics, students, teachers, borrowers, bystanders). It depends on the ability to develop lines of inquiry and conclusions, iteratively adding, correcting, and redirecting ideas as they emerge from theorizing, research, experimentation, and rebuttal. As we've seen, the very acts of writing and reading engage critical faculties unavailable to the unassisted mind. Without the means to think on the page and react to what others have written, there's no science.

"The invention of writing was a prerequisite for the development of philosophy and science in the ancient world," says historian David C. Lindberg in *The Beginnings of Western Science*. "The degree to which philosophy and science flourished in the ancient world was, to a very significant degree, a function of the efficiency of the system of writing (alphabetic writing having a great advantage over all of the alternatives) and the breadth of its diffusion among the people."

Lindberg traces the impulse to study nature into prehistory but notes that the slow turn toward genuine science began in Mesopotamia due to writing. The Babylonians developed mastery over astronomical phenomena beginning in the second millennium BCE, cataloging stars and planets, recording their movements, and eventually calculating eclipses and conjunctions—all while producing cuneiform almanacs, tables, and other documents to capture and transmit the relevant data.

While this information was originally used for such nonscientific aims as determining omens and portents, it was later employed in naturalistic astronomical observation—"the beginning," says Lindberg, "of computational astronomy." And only the beginning because Hellenistic astronomers later adapted these methods and even their data for their own efforts; working in the second century CE, for instance, Ptolemy of Alexandria employed eclipse data gathered in Babylonian records nine centuries earlier.

"As a result," says Lindberg, "the astronomical enterprise set out on a course that would culminate millennia later in the achievements of Nicolaus Copernicus, Johannes Kepler, and others." And Copernicus and Kepler were similarly indebted to another ancient source, Euclid of Alexandria.

Here's Looking at Euclid

Biographical details on Euclid are less than scanty, but he bridged the fourth and third centuries BCE and penned one of the most influential mathematical books in history. Having inherited a jumble of geometrical concepts, Euclid painstakingly edited them into four hundred propositions, sorted into thirteen books (what we'd call chapters), totaling twenty thousand lines of Greek, complete with diagrams.

Euclid's *Elements of Geometry* starts with the basics and progresses through layers of ascending conceptual complexity. Beginning with Book 1, readers first encounter points, lines, angles, triangles, and the Pythagorean theorem. By the time Book 13 rolls around, they're ready to tackle the Platonic solids—not only the standard cube and the hexahedron but also

- the tetrahedron, with its four triangular faces;
- the octahedron, with its eight triangular faces;

- the dodecahedron, with its twelve pentagonal faces; and
- the icosahedron, with its twenty triangular faces.

Along the way, they pick up a framework for parallel lines; the geometry of circles; the volume of pyramids, cones, cylinders, and spheres; number theory; and more.

While modern high school students might cringe at the work such exercises suggest (I certainly did) and encounter the Platonic solids only in the form of many-sided dice while playing Dungeons & Dragons, the ancients devoured Euclid's work and applied it to every discipline of thought imaginable. The mathematician, said historian Benjamin Wardhaugh, "arranged and exemplified a toolbox of the basic techniques and results that he had inherited: ways of arguing, ways of proving; facts that geometers commonly assumed or used, but seldom proved in full." Students, scholars, and practitioners alike could consult the *Elements* and employ its axioms, formulas, logic, and methods to yield reliable answers to their own quandaries.

To stress an earlier point, Euclid's *Elements* was not simply a conveyor of information; it was also a partner in novel thinking, fresh problem-solving, and the development of new models. The simple fact that the

Fragment from a first- or second-century copy of Euclid's *Elements*, including one of the earliest diagrams from the work. *SOURCE:* PAPYRUS OXYRHYNCHUS 29. FINE ART IMAGES/ HERITAGE IMAGE PARTNERSHIP LTD, ALAMY.

book survived all those centuries is remarkable, but that it continued to underwrite the imaginative powers of mathematicians is phenomenal—and that, of course, also explains why it survived, pushing its way along the z-axis century after century. Euclid's book applied to everything from astronomy to architecture, and manuscript evidence shows its readers engaged energetically with its concepts, scribbling in the margins as they commented on the axioms or worked them out for themselves.

Through the centuries, the *Elements* traveled from the libraries of Greek mathematicians to those of Byzantine statesmen, churchmen, and scholars, to the observatories of Muslim astronomers, and to medieval monasteries and universities across Europe as it became core to the curriculum. The book's presence could be felt not only in copies of itself but also woven into the work of later scholars. In the thirteenth century, for instance, University of Paris professor John of Sacrabosco used Euclid's *Elements* to create his own book of cosmology, *The Sphere*.

The *Elements* remained popular and useful into the age of print, when its intellectual footprint expanded even further. We've already seen the paltry number of books before printing. Even for an essential book like the *Elements*, it's possible that no more than a thousand copies existed across all of Europe, painstakingly copied and illustrated one at a time over hundreds of years. Working in his print shop in Venice, Erhard Ratdolt changed that in 1482 when he pressed out his first edition of the *Elements*, producing more than a thousand copies in just months.

And printers like Ratdolt were only getting started. Within another 120 years, various shops would produce forty new editions of the *Elements*. Tack on another century, and an additional three hundred would emerge.

In time there would be well over a thousand editions featuring extensive variation. Some featured reams of commentary; others truncated the text to its bare minimum. One might cover a thousand pages, another less than fifty. "In the world of print," says Wardhaugh, "Euclid's *Elements* was rapidly ceasing to be a single text, and becoming a wide tradition of different texts meant to be used in different ways by different kinds of people," including students, teachers, astronomers, sailors, artists, and others. The *Elements* was a portable tool for thinking through a variety of problems and challenges.

Page from Erhard Ratdolt's 1482 printed edition of Euclid's *Elements*, featuring diagrams and aftermarket marginalia. SOURCE: FOLGER SHAKESPEARE LIBRARY.

More Books, More Science—Sort Of

In conjunction with other contemporary forces, printing revolutionized science. The "mass mediatization" we encountered earlier bore immediate fruit in the many disciplines that constitute the enterprise: physics, astronomy, optics, chemistry, medicine, biology, and more. We can update David C. Lindberg's observation about writing: The degree to which science flourished in the modern world was, to a significant degree, a function of printing and the breadth of its diffusion among the people.

Historian William E. Burns offers several reasons why. First, printing helped stabilize texts. Whereas hand-copied manuscripts of the same book differed among themselves, sometimes radically, copies of printed editions basically matched no matter how many were produced. This consistency allowed scholars to use and reference the same texts, which aided shared understanding—or facilitated more beneficial disagreements.

Relatively stable texts also permitted scholars to use tools such as indexes more effectively. As we saw with Galen's codices, an index in a manuscript copy was useful but entirely self-referential; it worked only with that particular copy. The breakthrough came when scholars all over could use matching indexes in their various copies to compare and contrast ideas, not to mention navigate the rising deluge of information.

Second, as just alluded to, printing disseminated information. Most of the work in science involves the grueling collection of data. It's one of the reasons Darwin delayed publication for two decades; he didn't think he quite had enough evidence to make an irrefutable case, and so he kept building his argument, filling another notebook, penning another brief. Printing gave scientists access to previously collected data, which accelerated their own work.

A year after printing his Euclid, for instance, Ratdolt published an edition of the Alfonsine Tables, astronomical data collected in the thirteenth century and modified thereafter. The tables helped astronomers understand planetary motion and celestial events such as eclipses by using fixed stars to help compute the location of the sun, moon, and planets. Nicolaus Copernicus cherished his copy and used it to help formulate his heliocentric model of the universe. The Alfonsine Tables were also used in education and applied areas such as nautical navigation and calendar reform.

Third, printing illustrated its subjects. Ratdolt's *Elements* featured more than four hundred diagrams, allowing readers to visualize Euclid's otherwise abstract reasoning. By amplifying specificity and reducing ambiguity, illustrations helped all sorts of scientific inquiry, especially anatomy and the natural sciences, such as botany and zoology.

Readers valued the accurate, intricate, closely detailed woodcuts in Andreas Vesalius's 1543 treatise *On the Fabric of the Human Body* as much as the text. The artist who rendered them? Jan Stefan von Calcar, who studied under Renaissance master Titian. Von Calcar's striking imagery influenced medicine for generations.

Some of the leading artists of the period contributed to projects. Albrecht Dürer, for instance, engraved an image of a rhinoceros that found its way into popular zoological tomes, such as Conrad Gessner's

History of the Animals (1551) and Edward Topsell's *Historie of the Foure Footed Beasts* (1607), both replete with images of the creatures described.

Of course, the printing press also underwrote a lot of hokum. Printing did not ensure scientific veracity. Along with his Euclid and Alfonsine Tables and other legitimate scientific work, Ratdolt printed an edition of Duke Leopold of Austria's thirteenth-century guide to astrology, *Compilation of the Science of the Stars* (1489), which covered such details as how the stars influence weather, predict fortunes, and determine friendship, enmity, death, and much else. What we would regard as astronomy and astrology were overlapping disciplines at the time, and legitimate scientists pursued both. Galileo Galilei discovered the phases of Venus and the moons of Jupiter, cementing the case for a heliocentric universe; he also cast horoscopes.

A similar muddle persisted in the natural sciences. Little separated alchemy and legitimate chemistry, and alchemists used their supposed magic for all sorts of practical purposes. But eventually the practical won out over the esoteric, and concern for discovering the philosopher's stone or mixing the elixir of life faded while more naturalistic matters pushed to the fore.

Publishing history reveals the transition. When, for instance, Frankfurt printer Christian Egenolff produced his German-language manual *The Right Purpose of Alchemy*, he based it on an existing manuscript by alchemist Petrus Kertzenmacher discussing, among other things, the supposed spirits of metals and methods for transmuting lead into gold. But Egenolff excised the magic from his edition, transmuting it into a plain and useful technical manual for metalworkers and jewelers.

A similar transformation was underway in zoology. Edward Topsell's *Historie of the Foure Footed Beasts* featured several fantastical animals, such as the unicorn and manticore, alongside its dog, cat, cow, goat, and rhinoceros. But Topsell's "monumental work," as University of Houston professor John Lienhard says, "was actually an early glimmer of modern science. For all its imperfection, it represents a vast collection of would-be observational data, and it even includes a rudimentary rule for sifting truth from supposition."

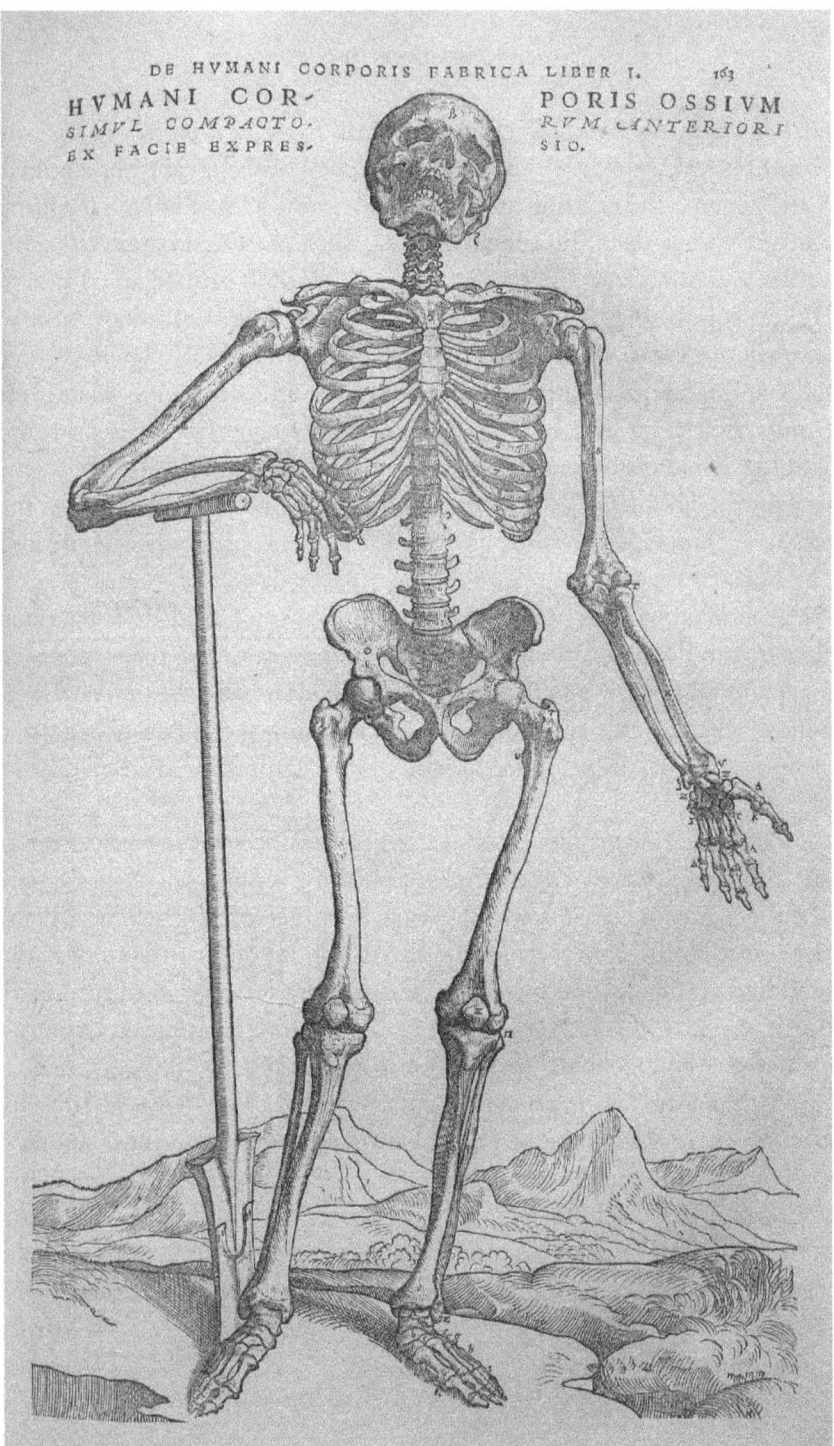

Digging your own grave? Image from Andreas Vesalius's *On the Fabric of the Human Body*. SOURCE: WIKIMEDIA COMMONS.

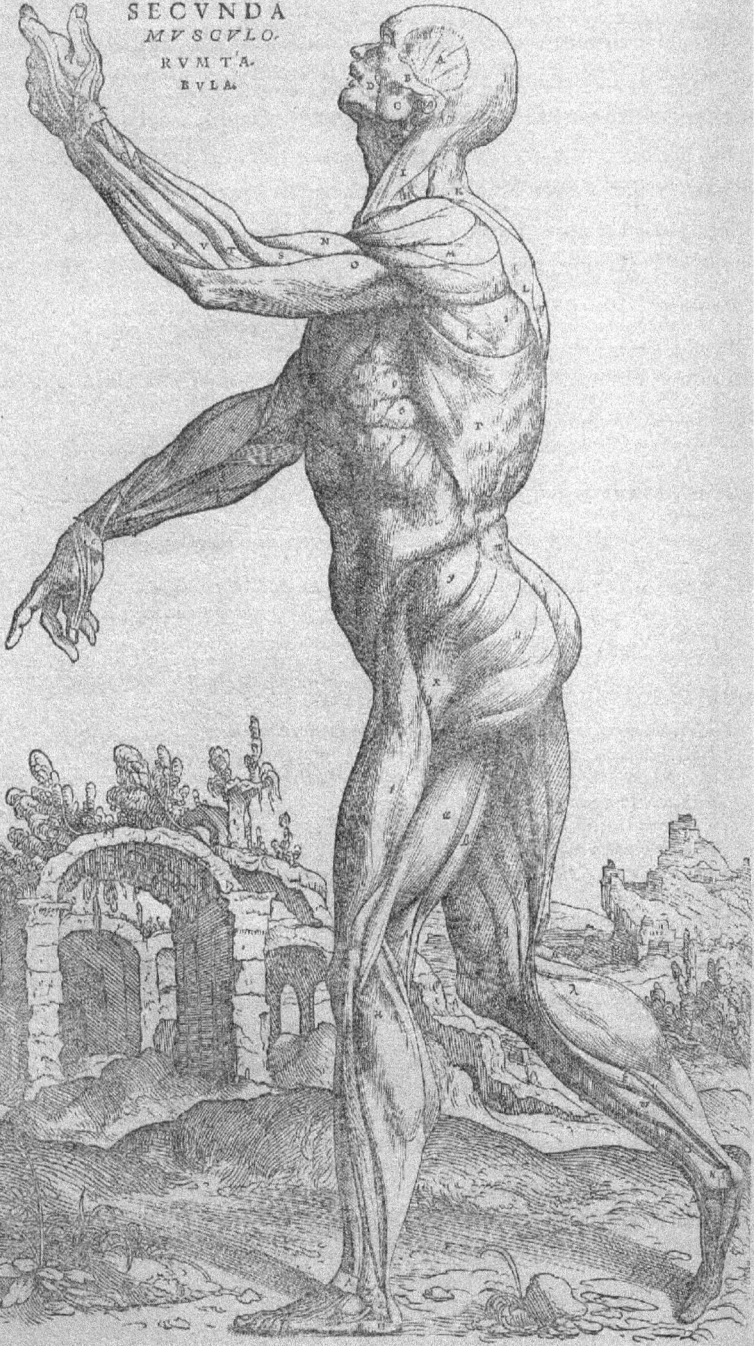

Muscle man. Image from Andreas Vesalius's *On the Fabric of the Human Body*. *SOURCE*: WIKIMEDIA COMMONS.

While printers spread bogus notions and increased the likelihood that some would accept utter nonsense, they also created the conditions under which those ideas could be corrected. A suppressed idea can spread on the sly, immune to correction; meanwhile, an expressed idea can be refuted. Printing accelerated this process by exposing the errors to more readers with competing observations and theories, and it did so at scale. Because books are tools for active thinking, not merely containers of information for passive consumption, even erroneous work can serve constructive ends. As far as the development of science is concerned, publishing errors is a necessary part of advancing knowledge, helping refine what we accept as real and true.

Rumor Has It

Most of the nonsense contained in these books persisted by means of cultural inertia. While the humanists were eager to readopt the classics, they also readopted their epistemological shortcomings. Ancient and medieval science tended to focus on written sources—that is, books that could be consulted for supposedly settled truths.

In the preface to his *Natural History*, for instance, Roman naturalist Pliny the Elder explains, "I have included . . . 20,000 topics, all worthy of attention . . . gained by the perusal of about 2,000 volumes, of which a few only are in the hands of the studious, on account of the obscurity of the subjects, procured by the careful perusal of 100 select authors." Nor was this all. "To these," Pliny says, "I have made considerable additions of things, which were either not known to my predecessors, or which have been lately discovered." But those late discoveries weren't verified with any more rigor than the rest of what he compiled. For Pliny, science was gathering data from prior publications and filling out the portrait with colorful rumor.

Aristotle followed a similar method, though he was less gullible than Pliny and supplemented his literary scouring with genuine, get-your-hands-dirty investigation. In his books on biology, for instance, Aristotle passed along myths and errors, but he also described the internal anatomy of more than a hundred animals. "For about thirty-five of them," as

Armand Marie Leroi notes in *The Lagoon*, "his information is so extensive or accurate that he must have dissected them himself."

Still, Aristotle was an outlier. The trend throughout the medieval era was to parrot prior authorities. Conrad Gessner's background as both a bibliographer and a naturalist underscores this point; in *History of the Animals*, he aimed to compile everything ever written about a whole host of creatures. Given such heavy reliance on prior opinion, nonsense like we've already seen could easily creep into the pages.

Without the ability to directly observe an elephant or, say, an armadillo, Gessner was forced to rely on reports. "This is why," says Cambridge fellow Sachiko Kusukawa, "the elephant, the armadillo and the unicorn are given equal credence." Gessner found references to unicorns in the Bible, Aristotle, Pliny, and other sources and so passed along the information as credible. Like Pliny, Gessner relied on revered sources as a methodological preference; also like Pliny, he employed contemporary reports, though with similar critical rigor (which is to say, less than modern practitioners would allow).

Even when Gessner could rely on direct observation of the common guinea pig, for instance, he defaulted to reports. "It was a fundamentally bookish way of forming knowledge," says Kusukawa. It was also coming to an end.

New World, New Science

What changed this limited trajectory of Western science? Four interlocking developments. First came an encounter with the New World. After Christopher Columbus initiated contact with the Americas, Europeans faced flora and fauna about which their authorities were mute. The New World teemed with species never encountered in the classical sources. An authoritative ancient collection of herbal treatments by Dioscorides, for instance, covered five hundred plants. But when Francisco Hernandez, physician to King Philip II of Spain, documented the natural history of the king's new lands in 1570, he listed six times that number.

Second, and closely related to the first, was empiricism. Instead of reshuffling known items and preexisting categories, European naturalists directly experienced the novelties they encountered. And direct observation only increased as discovery intensified. "At the beginning of the seventeenth century, European naturalists had identified around 6,000 different species of plant," says historian James Poskett. "By the end of the eighteenth century, they had identified over 50,000 species, the majority of which originated outside of Europe."

These encounters not only revealed the insufficiency and ignorance of classical authors but also exposed their errors. Ancient authors reported rumors of fantastical humans of wild shapes and habits in foreign lands. In his *Natural History*, for instance, Pliny mentions "the Nigroæ, whose king has only one eye, and that in the forehead, the Agriophagi, who live principally on the flesh of panthers and lions, the Pamphagi, who will eat anything, the Anthropophagi, who live on human flesh, the Cynamolgi, a people with the heads of dogs, the Artabatitæ, who have four feet, and wander about after the manner of wild beasts."

Not everyone accepted such reports. Augustine had his doubts. "We are not bound to believe all we hear of these monstrosities," he said. Nevertheless, medievals inherited these and similar accounts, popularizing them through books such as the *Alexander Romance* and *Livre des Merveilles du Monde*, a fifteenth-century French exploration of supposed natural wonders. In islands off the coast of India, Marco Polo claimed to have seen people with "heads like dogs, and teeth and eyes likewise; in fact, in the face they are all just like big mastiff dogs!"

Explorers found wonders of all sorts, but nothing like what the classical or medieval authors suggested they might encounter. Christopher Columbus kept, read, and annotated his copy of *The Travels of Marco Polo* but noted the absence of dog-headed denizens of what he assumed was India. "I have not found the human monsters which many people expected," he wrote in his first letter to Spain. "On the contrary, the whole population is very well made."

What's more, according to Aristotle, none of those beautiful people should have been there at all. The Caribbean lies in the so-called torrid zone, which the philosopher and those following his lead assumed was too hot to support human habitation.

Time to revise! But how? Scientists adopted an inductive, empirical approach, championed by people such as Francis Bacon in his *Novum Organum.* "The solution to the sterility of ancient science," says Burns, "was to reroot science in empirical reality rather than ancient or modern theory, and build new theories on a base created by observation."

To these first two developments we can add a third—a newfound and lasting reliance on purely naturalistic explanations. Anaximander's preference for naturalism had long been the minority view. Of course, it did have sympathizers and supporters, none more influential than Epicurus in the third century BCE and his acolyte Lucretius in the first century BCE.

While the view waned in the Middle Ages, Christian monks found Lucretius's poem *On the Nature of Things* valuable enough to preserve it through the Middle Ages, despite having a complicated relationship with its underlying philosophy. Its real influence came when Poggio Bracciolini rediscovered a copy in the early fifteenth century. Once celebrated and circulated by Renaissance humanists, Lucretius's naturalistic worldview would go on to influence everyone from Galileo to René Descartes, Isaac Newton, and far beyond.

Ancient and medieval scholars did offer naturalistic explanations for phenomena, though unevenly and inconsistently. But at the start of the modern era, naturalism went from an occasional viewpoint to a methodological priority. The commitment to empirical studies and naturalistic explanations for observed phenomena made it possible to generalize about causes and other relationships, allowing for the formulation of provable theories and natural laws. And while scientists might personally confess belief in God, the miraculous, and so on—many did, in fact—they separated those beliefs from the practice of research and reporting their findings.

Which leads to the fourth and final development.

Essential Tool of Discovery

Just as it's impossible to imagine science without books, it's likewise impossible to imagine the scientific revolution without printing. If we resume exploring William Burns's list of the benefits that printing brought to science, this is surely among the biggest: Printing spread fresh data, calculations, information, theories, and more. As new discoveries and explanations began bubbling up, they were captured in letters, books, and articles, which then spread abroad and could be used by others. And since those books were full of reports of empirical, naturalistic analysis, their results were reliable enough to inform the work of further scientists.

Historian David Wootton includes printing as a key ingredient of the scientific revolution. Not only did scientists have new tools, such as telescopes, microscopes, and barometers, in this period, he says, but "they [also] had the printing press . . . which created new types of intellectual community and transformed access to information." In just the first hundred years following the advent of the printing press, at least nine hundred individual first editions of scientific interest appeared across Europe, according to a bibliographic study by Margaret Bingham Stillwell:

- Medicine: 304 individual editions
- Natural science: 158 individual editions
- Mathematics: 130 individual editions
- Astronomy: 124 individual editions
- Technology: 98 individual editions
- Physics: 86 individual editions

Printing created a positive feedback loop in which each new book served to prompt the publication of additional editions, each seeking to advance its respective field. The preeminent names of the period are preeminent for a reason. Copernicus, Kepler, Galileo, Newton—we know their names because they advanced their areas of interest through discoveries and models disseminated in their writing and the writing of those who further developed their insights.

Printing enabled the cumulative, iterative aspects of science to happen at scale. Science is incremental. It grows bit by bit. One observation serves as the foundation for the next. This subsequent work is usually advanced by someone other than the original inquirer. Copernicus depended on the Alfonsine Tables but contradicted and surpassed them. Galileo likewise depended on Copernicus but contradicted and surpassed him. The nature of science is to develop workable models that become ascendant until unseated by better models. And printed books were the means by which those models could both rise and fall. Printing became so intertwined in the scientific enterprise that astronomer Tycho Brahe, as Burns notes, installed a press in his observatory and even purchased a paper mill.

The ingredients of future discovery spread through networks of personal interaction, letters, papers, and books. Even with limited circulation throughout the classical and medieval period, books proved essential. But the scale of information enabled by printing altered the course of history, something easily seen in individual cases, such as that of Charles Darwin.

Origin of the *Origin*

Darwin studied the visible world, but he was trying to understand something far from apparent. Following his return from a scientific expedition to the Galápagos Islands on the HMS *Beagle*, he'd become convinced of the mutability of species, that one shaded into another and that ultimately life had descended from a common ancestor. But how?

He began keeping a notebook, jotting down thoughts, observations, conjectures, and conclusions. When that notebook was full, he began another. Then another. And another. The very same ones returned in the pink gift bag after their temporary pilferage. As the pages turned, his theory evolved. Darwin's notebooks were like the fossil record of its growth and development.

He eventually sat down and wrote a summary of around three dozen pages—then, some time after that, a longer, more developed version of the theory. But he kept brooding and noodling, experimenting and explaining, all the while adding facts and refining the theory. He was finally forced by circumstance to write the book that became *On the Origin of Species*. Another scientist had developed nearly the exact same theory; if Darwin didn't want to get scooped, he had to publish. Thus it finally came after twenty years.

Importantly, Darwin's *Origin* contains a fossil record of its own: the intellectual and theoretical framework of the project itself. Along with his own observations, Darwin depended on the prior research of others to build his case. "Darwin," says University of Florida historian Vassiliki Betty Smocovitis, "relied on mathematical calculations based on a series of floras compiled by botanists like Joseph Hooker (and his flora of New Zealand), Asa Gray (and his celebrated flora of temperate North America), Hewett C. Watson, and others to lay the groundwork for his belief that varieties were indeed incipient species."

Working with previously published studies, such as Gray's "Statistics of the Flora of the Northern United States" and Alphonse de Candolle's *Géographie Biologique Raisonnée*, Darwin used mathematical analysis to establish his case. "Though he was not mathematically inclined," says Smocovitis, "Darwin was aware of these studies and looked to them to provide numerical evidence of a correlation between extensiveness of plant distribution and variability."

But this was true beyond the nitty-gritty, micro level of proving the theory. It was also true at the airy, lofty, macro level of the theory itself. Darwin's thinking was prompted by such books as the anonymously published *Vestiges of the Natural History of Creation* (whose author was later identified as Robert Chambers), Charles Lyell's *Principles of Geology*, and Thomas Malthus's *Essay on the Principle of Population*, without which there would have been no *Origin of Species* because it supplied Darwin with the idea of natural selection.

How important were these books and published studies? The scientist who nearly scooped Darwin was Alfred Russell Wallace, working half a world away in the Dutch East Indies. Wallace had read Darwin's book on the *Beagle* voyage, but, as Darwin biographer Janet Browne notes, Wallace had also read *Vestiges* and Malthus and ultimately come to the same conclusions Darwin did. The ready circulation of ideas in the post-Gutenberg era facilitated discovery and the expanding knowledge of the physical world. As we'll see in the following chapters, it facilitated much else besides.

Marginalia: The Third Variable

When we think of science and religion, we often imagine bitter foes—Montagues and Capulets locked in a never-ending, fundamental antagonism. But the historical record offers us a different and more interesting story.

Both the Muslim East and the Christian West saw significant scientific achievements flourish within overtly religious cultures. Between the eighth and thirteenth centuries, during the so-called Islamic Golden Age, Muslim scholars produced pathbreaking work in mathematics, medicine, chemistry, optics, and astronomy. And later, in the Christian West, scientists like Copernicus, Galileo, and Newton viewed their work as uncovering the laws of a rational, orderly universe created by God.

Belief and science are not at odds, or at least they need not be. Yet science ran aground in the East. While the Islamic East was scientifically far more advanced than the Christian West for centuries, a radical split began in the fifteenth century. Why? To get a fuller picture, we must look beyond science and religion alone. And, conveniently enough, these rival cultures present us with a test case for a third variable—that is, receptivity to technological change, in this case printing, which directly impacted the production and use of scientific information.

As we've seen, Islamic authorities banned the use of printing for Muslims. This decision affected not only religious texts but also scientific texts. As a result, the spread of scientific information was limited to the speed and scale of individual scribes dipping pens in their inkwells.

Meanwhile, in the West presses poured out pages faster than people could read them. Scientific works, such as Copernicus's *On the Revolutions of the Heavenly Spheres* and Galileo's *Dialogue Concerning the Two Chief World Systems*, were widely distributed, allowing scientists to study, critique, replicate, and develop each other's work. By enabling the mass production of books and journals, the press accelerated the sharing of ideas, widened their dissemination, and encouraged fruitful intellectual collaboration and competition.

In the framework of Virginia Postrel, a journalist known for her analyses of economics, technology, and cultural innovation, the Christian West embraced a *dynamist* position toward the printing press, while the Islamic East embraced a *stasist* position. Dynamism, as Postrel explains, is characterized by openness to experimentation, innovation, and change. Stasism, by contrast, emphasizes caution, preservation, tradition.

Europe's rapid adoption of the printing press exemplifies Postrel's dynamist mindset. Though the Catholic Church attempted to regulate printing and restrict some scientific texts, it wasn't terribly successful, and Protestant nations allowed greater freedom. As a result, scientific information proliferated and quickly fed its own expansion.

The opposition to printing in Islamic lands, however, exemplifies Postrel's stasist mindset. Concerns about the sanctity of religious texts and worry about undermining religious authority—both legitimate fears—delayed the adoption of the printing press and contributed to a marginalization of Islamic science. While science exploded in the West, it retreated in the East; relying on handwritten manuscripts bottlenecked the dissemination of information and limited the ability of Muslim scientists to participate in the growing conversation.

Consider some of the key Muslim scholars who impacted Western thought:

- Al-Khwarizmi (c. 780–850) influenced Western mathematics; we can thank him for the terms *algebra* and *algorithm*.
- Al-Razi (c. 854–925), also known as Rhazes, influenced Western medicine and chemistry.
- Ibn Sina (980–1037), also known as Avicenna, influenced Western medicine and philosophy.
- Al-Haytham (c. 965–1040), also known as Alhazen, influenced the Western understanding of optics.
- Nasir al-Din al-Tusi (1201–1274) influenced Western mathematics and astronomy; American high school students can probably blame him for trigonometry.

These contributions were essential to the flowering of intellectual life in Europe. But note the lifespans: The majority of important Islamic contributions to Western scientific thought occurred by the end of the thirteenth century.

While Islamic scientists continued to work and make contributions in subsequent centuries, their influence waned, even in their own countries. Mongol incursions and conquest, political upheaval, and shifting cultural priorities all played a part in this process. But it seems impossible to overrate the role suppressing the printing press played in stifling Islamic science; that, more than any fundamental properties of the Islamic religion itself, seems to have been a deciding factor. A dynamist approach to the technology of printing would have likely produced a far different outcome.

Chapter 12

FOUNDED ON BOOKS

"Bibliomany Has Possessed Me"

In the summer of 1775, while traveling to Philadelphia to serve in the Second Continental Congress, a thirty-one-year-old Thomas Jefferson stopped off in Annapolis, Maryland, to shop for books. He did the same thing when he later arrived in Philadelphia and again before he left for home in August. Jefferson had several vices. Profligacy with books was one.

He admitted as much in letters. Thanking a friend who was purchasing books for him, Jefferson explained his aversion to high-priced volumes. "Sensible that I labour grievously under the malady of Bibliomanie," he wrote from Paris in 1789, "I submit to the rule of buying only at reasonable prices, as to a regimen necessary in that disease." But he wasn't as sensible as he might sound here. Jefferson incurred vast amounts of debt purchasing books.

His borrowing began small. In 1783, for instance, he bummed $133 from James Madison (around $4,000 in 2024 dollars) in part to buy books. But it snowballed. Because of inherited debts and his refusal to curb his purchasing of books, art, clothes, and wine, Jefferson was constantly in arrears, having to negotiate with lenders and restructure his debts. Creditors dogged him to the end of his life.

Even while showing restraint, he betrayed his profligacy. When Thomas Law asked whether Jefferson might subscribe to a new book—subscription being the eighteenth- and nineteenth-century version of Kickstarter or GoFundMe—Jefferson declined. "I am now entered on

my 69th year," he answered in his 1811 response, continuing, "The tables of mortality tell me I have 7 years to live. My bibliomany has possessed me of perhaps 20,000 volumes. Of these there are probably 1,000 which I would read. . . . But it is also probable I shall decamp before I get through 50 of them. Why then add an unit to the 19,950 which I shall never read? . . . Sober reason tells me it is time to leave off buying books."

Twenty thousand volumes is an astounding number of books for a personal library, especially in the nineteenth century. It was also wildly overstated. Still, who was he kidding? Jefferson was nowhere near done buying books. Sober reason was no match for his bibliomania, and, as he wouldn't die until 1826, he had more than twice as many years remaining as those actuarial tables suggested—plenty of time to purchase more.

"Rapid Progress of Knowledge"

Jefferson represents an extreme form of a common type among the founding generation: the intellectually curious whose knowledge of the world was informed by untold gallons of ink. "The Americans were," says historian Forrest McDonald, "a remarkably literate people." Almost everyone had access to the news. "It has been estimated that newspapers went into roughly 40,000 homes on the eve of the Revolution, and possibly twice that number by the end of the century," he says. That's a tenth of homes and doesn't account for the postpurchase spread of broadsheets and pamphlets through towns or their circulation at taverns, inns, and coffee shops. By 1790, more than sixty different locales boasted newspapers, producing some ninety-nine publications in all, many printing thousands of copies every week. By the end of Jefferson's presidency, the numbers had more than tripled.

Much of this news and pamphleteering was highly partisan, increasingly so over the decades, and warranted its own criticism. In 1834, novelist James Fenimore Cooper described the American press as "hurtful," "corrupting," and spreading "an atmosphere of falsehoods." Modern

Thomas Jefferson, dreaming about his next book purchase. Portrait by Mather Brown. *SOURCE:* NATIONAL PORTRAIT GALLERY. WIKIMEDIA COMMONS.

fretting over bias or fake news had nothing on the eighteenth and nineteenth centuries. But then readers were not limited to the news.

McDonald quotes a 1766 edition of one of those infamous papers, the *New York Gazette and Mercury*, enthusing about the expanding book trade: "Every lover of his country hath long observed with sacred pleasure, the rapid progress of knowledge in this once howling wilderness, occasioned by the vast importation of books; the many public and private libraries in all parts of the country; the great taste for reading which prevails among people of every rank." Imports tell only part of the story, of course. Domestic printing grew in the period as well. By 1771, for instance, Boston boasted

- ten printer-booksellers
- eight bookbinders
- seven book importers
- seven "firms in the book trade"
- four printers
- two stationers

Still, book stocks were meager, especially outside major cities. Following the example of Ben Franklin in Philadelphia, lending libraries sprang up throughout the colonies, but libraries in country towns might sport fewer than a hundred titles.

The scarcity of books didn't stop Americans from reading. What books passed beneath their eyes? The Bible was ubiquitous, as were religious books of all sorts and certain classics, especially Shakespeare. When traveling through America in the early 1830s, Alexis de Tocqueville said the Bard could even be found in the huts of frontiersmen. "I remember reading the feudal drama *Henry V* for the first time in a log cabin," he said. In one visit to a frontier cabin, he spied "a shelf of rough-hewn lumber" with "several volumes . . . a Bible, the first six cantos of Milton, and two plays by Shakespeare."

McDonald notes a special love for the classics and history—such as Vergil, Cicero, Tacitus, and Thucydides—sometimes in the original

languages but usually in translation or in abridgments and populariza-
tions. "Charles Rollin's two-volume *The Ancient History*, an abridgment in
translation of Greek and Latin authorities, was widely read in America,"
he says, "as were David Langhorne's edition of *Plutarch's Lives* and James
Hampton's 1762 translation of *The General History of Polybius*, which
went through four editions." Contemporary historians writing about the
ancient world enjoyed similar popularity.

British history proved even more prominent than that of ancient
Rome and Greece. "Our laws, language, religion, politics and manners are
so deeply laid in English foundations, that we shall never cease to consider their history as a part of ours," as Jefferson said when praising David
Hume's *History of England*. Americans used books of and about the past
to orient themselves in the present, especially regarding their relationship
to England and when founding their new nation. Books revealed patterns, provided trains of thought, supplied arguments and evidence, and
engendered a feeling of the world as rational (or at least understandable).
For those who recognized the value of books, building a library was a
meaningful activity. For those who depended on books for their involvement in government, it was essential.

Two Trunks of Books

Jefferson and James Madison first met in 1776 and, while working
together, became fast friends. The two possessed similar outlooks on life,
both politically and philosophically, and shared an equal curiosity about
natural science. With such mutualities, it's no surprise books formed a
running theme in their relationship.

In 1782, Madison found himself responsible for drawing up a catalog of titles "proper for the use of Congress." As debate and deliberation
might require information and insights stretching far beyond the ken
of even the most eminent and notable citizens of the young nation,
the legislative body needed a library to supply facts, precedents, and
arguments—the raw, intellectual materials for informed proposals and
decisions. Madison drew on all resources available to conjure a useful
list, not only his own reading but also the libraries of family, friends, and

associates, along with the shelves of various institutional libraries and those of booksellers he frequented.

While Madison and Jefferson roomed at the same Philadelphia boardinghouse the following year, the pair conferred on the project. When it was finished, Madison's list contained more than five hundred titles in roughly thirteen hundred volumes and included selections on law and legal theory, general history, various national histories, collections of treaties, diplomacy, geography, and more. Though Madison's work was largely done at that point, his final list does show parallels to a similar catalog Jefferson made that year for his own use and reflects their conversations on the work.

Later, as Jefferson prepared to leave for Paris as ambassador to France, he wrote Madison from Annapolis in May 1784 to let him know that he would take advantage of the position to help his friend build his personal library. "In the purchase of books, pamphlets &c. old and curious, or new and useful I shall ever keep you in my eye," said Jefferson. Humorously, at the bottom of this letter Jefferson mentions $503 he still owes Madison, along with repayment of $407 toward the total.

Following the pair's letters for the next year and several months yields an unfolding drama. What books? When will they come? To what use will Madison put them? Based on their boardinghouse conversations about books, Madison's interest and anticipation must have been piqued.

Once in Paris, Jefferson frequented bookstalls and bookshops, keeping his eyes peeled as he suggested. By November he'd acquired several choice volumes and wrote Madison to let him know. "I shall subjoin the few books I have ventured to buy for you," he said, listing Juan de Mariana's *Histories of the Affairs of Spain*, Jean-Zacharie Paradis de Raymondis's *A Treatise on Morality and Happiness*, a book on diplomacy by Dutch diplomat Abraham de Wicquefort, and a five-volume French–Latin dictionary that served as a source for Denis Diderot and Jean le Rond d'Alembert's famous *Encyclopédie*.

The original *Encyclopédie* had recently been expanded by an enterprising publisher; Jefferson mentioned thirty-seven volumes and said he would subscribe and send along what volumes he could. "I have been induced to do it by the combined circumstances of their utility and cheapness," Jefferson explained. But cheap and useful could guide only so far. "I wish I had a catalogue of the books you would be willing to buy, because they are often to be met with on stalls very cheap, and I would get them as occasions should arise."

The pair wrote back and forth about goings on in the Virginia Assembly and Congress, along with news of the European nations and references to shenanigans even further afield, such as the capture of the American ship *Betsy* by the Barbary pirates. But there in the background hummed the question of the books. "I mentioned to you in a former letter a very good dictionary of universal law called the Code d'humanité in 13. vols.," wrote Jefferson in March 1785. "Meeting by chance an opportunity of buying a copy, new, and well bound for 104 livres I purchased it for you."

The books were stacking up, but when would he send them? Jefferson said he thought he might send them as early as May but didn't know where Madison would be, hearing that his friend had been nominated for a post in Spain. In lieu of sending along the whole lot, he sent a token: two pamphlets, one on animal magnetism and another dealing with an aeronautical expedition—that interest in science coming to the fore.

Madison was pleased to receive them for their own sake and the larger pile they represented. "I thank you much for your attention to my literary wants," he wrote Jefferson in April. "All the purchases you have made for me, are such as I should have made for myself with the same opportunities."

Responding to Jefferson's request for a shopping list, Madison demurred. "I am afraid if I were to attempt a catalogue of my wants I should not only trouble you beyond measure, but exceed the limits which other considerations ought to prescribe to me," he said. But that didn't stop him from rattling off a list nonetheless. You can almost feel his anticipation build as he adds to the tally.

Madison was especially keen for "treatises on the antient or modern fœderal republics, on the law of Nations, and the history natural and political of the New World." To those he added several others, including French tracts on various national economies, rare Greek and Roman authors "where they can be got very cheap," historians of the Roman

James Madison, books in the background. Portrait by Gilbert Stuart.
SOURCE: ZURI SWIMMER, ALAMY.

Empire in decline, Blaise Pascal's *Provincial Letters*, and an original Spanish edition of *A Voyage to South America* by scientist and seaman Antonio de Ulloa. Madison's interest in science and the natural world remained high, and he asked Jefferson for the "best edition" of an unspecified book by Carl Linnaeus, any new books by naturalist Georges-Louis Leclerc, Comte de Buffon, and a recently published travelogue of China.

Jefferson was prevented from making a lengthy reply or sending more information on his book buying but said in May that he'd send the books on the next ship out of Le Havre, a port in Normandy. But when would that be? Madison was getting antsy: "I have not yet received any of the books which you have been so kind as to pick up for me, but expect their arrival daily," he wrote in August.

A month later, Jefferson finally sent the books, along with an accounting of the prices paid for each volume. He provided books of politics, history, law, economics, philosophy, and science, many for which Madison had specifically asked and many that Jefferson simply believed his friend would find useful—dozens of volumes, enough to fill two trunks—including his Buffon and several volumes of Linnaeus.

It could have been more. "I have at length made up the purchase of books for you, as far as it can be done for the present," he said, lamenting titles he couldn't turn up. No matter. Some of the unfindables Jefferson knew could be purchased in England and mentioned dispatching someone to acquire them. In the meantime, a treasure sat in the cargo hold of a ship bound for America and for Madison's library.

During his stay in Paris, Jefferson picked up more than a few volumes for his own library as well. The tally ran to 1,850 books. "Upon returning to America in 1789," says Mark Dimunation of the Library of Congress, "Jefferson possessed a library twice the size of the one he owned at his departure and with"—this detail will become more important as the story progresses—"a considerable debt to show for it."

In Service of an Argument

Travel by ocean was slow. Madison didn't get his books until the new year. But when he did, he dove into the trunks. "Since I have been at home I have had leisure to review the literary cargo for which I am so much indebted to your friendship," Madison wrote in March. "The collection is perfectly to my mind." He knew exactly what he would do with it.

Following the Revolutionary War, Virginia, Massachusetts, Pennsylvania, and the other former colonies combined as thirteen sovereign states under the terms of the Articles of Confederation. This arrangement posed problems from the start, according to its critics. The individual states retained too much power at the expense of the national government. The Congress could not, for instance, levy taxes, regulate commerce, or enforce treaties. And while the states retained so much power, no national judiciary existed to handle disputes between them.

Madison was not a fan. He believed these drawbacks prevented the federal government from proper governance. The solution? A stronger, more centralized government. But how to make the argument, especially given his contemporaries' understandable skittishness about centralized power? And what shape should this new government take? Madison would turn to his trunks full of books for help in thinking through the answers to those questions and convincing his fellow Americans of his proposed solutions.

From April through probably June 1786, Madison scoured the volumes in his trunks for insights about confederations, their strengths and weaknesses, leaving behind a forty-two-page report of his findings. "Notes on Ancient and Modern Confederacies" documents the pluses and minuses of six separate governments, three ancient (Lycian, Amphictyonic, and Achaean) and three modern (Helvetic, Belgic, and Germanic). His detailed notes cover how power was shared and operated, how policies were enacted, and, most important, how the purposes of the various governments were frustrated by the design of those same governments. George Washington found the notes so helpful that he cribbed a copy, excluding the Latin.

An example from Madison's dissection of the Belgic confederacy:

> *It is clear that the delay occasioned by recurring to seven independent provinces including about 52 voting Cities &c. is a vice in the Belgic Republic which exposes it to the most fatal inconveniences. Accordingly the fathers of their Country have endeavored to remedy it in the extraordinary Assemblies of the States Genl. in [1584] 1651, 1716, 1717, but unhappily without effect. . . . Among other evils it gives foreign ministers the means of arresting the most important deliberations by gaining a single province or City. This was done by France in 1726, when the Treaty of Hanover was delayed a whole year.*

Along the way, he leaves a crumb trail of citations, indicating from which books he drew the fact or observation that undergirds his analysis, including Fortuné-Barthélémy de Félice's thirteen-volume *Code of Humanity*, the updated *Encyclopédie*, William Temple's *Observations upon the United Provinces of the Netherlands*, and eighteen other reference works.

Per the earlier discussions of the Attic Poet, Vergil, and Ovid, these notes don't simply reflect or record Madison's thinking. They are in his own handwriting; the reading and the writing are part of his thinking, which is happening in conversation with book after book. Reading "Notes on Ancient and Modern Confederacies" is watching a stage in the birth of the US Constitution. These notes would inform his later work—not only documents such as his "Vices of the Political System of the United States" and subsequent contributions to the *Federalist Papers* but also the live debates at the Constitutional Convention for which he was better prepared than any other participant.

"Every Person seems to acknowledge his greatness," said William Pierce, who attended the convention. "In a very able and ingenious Speech," he said, Madison "ran through the whole Scheme of the Government—pointed out all the beauties and defects of ancient Republics; compared their situation with ours wherever it appeared to bear any analogy."

Despite disagreement, convention-goers recognized his unique contribution. "He blends together the profound politician, with the Scholar," said Pierce. "In the management of every great question he evidently took the lead in the Convention. . . . He always comes forward the best informed Man of any point in debate. The affairs of the United States, he perhaps, has the most correct knowledge of, of any Man in the Union." It's reductionistic to say, but Madison possessed better, more compelling thoughts about national governance because he thought with the help of better, more compelling books. The same dynamic served his friend, as one particular episode reveals.

Jefferson Makes His Case

When the US government caught wind that France had declared war on Great Britain and the Netherlands in 1793, it caused a ruckus in President

George Washington's cabinet. How could a war across the Atlantic have any bearing on life in America? The United States had treaties with France that might oblige the young nation to defend French interests in the Caribbean against the British and invite entanglement with French privateers at British expense.

Since America's wealth was at that point heavily dependent on favorable trade with Britain, Secretary of the Treasury Alexander Hamilton favored canceling the treaties. But, as secretary of state, Jefferson pumped the brakes. Jefferson's approach to resolving the controversy presents an object lesson for several themes we've covered so far in this book.

On April 18, Washington asked his cabinet members to respond to a list of thirteen questions inspired by Hamilton regarding the treaties and their ongoing validity. He gave the men the evening to consider their responses before a 9 p.m. meeting at his residence. The National Archives preserves several documents that show the development of Jefferson's stance. As with Madison's "Notes on Ancient and Modern Confederacies," these documents not only record Jefferson's thinking but in part *are* his thinking, revealing the cognitive process in action.

In the first of three key documents, written after the initial cabinet meeting, Jefferson simply jots down his rough thoughts, beginning with a couple of principles that help govern the subsequent thinking. For Jefferson, the first principle—that "the people [are] the source of all authority" and "the Constituent in all treaties"—answers a handful of the president's questions in one fell swoop. His second, that "the Legislature alone can declare war," helps answer most of the remainder. But this wasn't enough to win the argument with Hamilton.

At the April 19 meeting at Washington's, Hamilton cited a passage from Swiss jurist and philosopher Emer de Vattel's book *The Law of Nations*. According to Forrest McDonald, Hamilton was regarded as "the nation's most learned expert on international law." He knew his way around all the principal authorities: Vattel, Hugo Grotius, Samuel Pufendorf, Jean-Jacques Burlamaqui, and Christian Wolff. But Jefferson had a copy of Vattel as well; in fact, he'd read all the same books as Hamilton and considered him totally off the mark.

"Would you suppose it possible that it should have been seriously proposed to declare our treaties with France void on the authority of an ill-understood scrap in Vattel . . . and that it should be necessary to discuss it?" he asked Madison incredulously in a letter sent amid the controversy. In answer to Washington and Hamilton, Jefferson consulted his copy and jotted down useful references in his notes, such as citations relating to "the validity of treaties," "nullity of treaties ruinous to a state," and "the viol[atio]n of a treaty is an injury."

He continued thinking on paper in a separate note, describing the risk of pulling out of the treaty "without just cause or compens[atio]n." It would give "cause of war to France." Along with that observation, he listed twelve questions of his own concerning the rationale for entering a war, the likelihood of obligations stemming from guarantees, the readiness and strategic interests, and the possible consequences of both action and inaction. Importantly, Jefferson also put the idea machine to work, listing several authorities beyond Vattel to weigh against Hamilton. He chose Hamilton's favorites: Grotius, Pufendorf, and Wolff.

Next, Jefferson drafted a nearly five-thousand-word memo to Washington on the matter. He closed with eight individual conclusions, emphasizing the continued validity and strategic relevance of the treaties, the limited or manageable costs of upholding the agreements, and the importance of maintaining neutrality between France and Britain.

As part of his case, he personally translated and quoted passages from Grotius, Pufendorf, Wolff, and Vattel in parallel columns for Washington, demonstrating the wrong in reneging on America's agreements with France, despite whatever arguments might support canceling the treaties. (He even had a snippet of Burlamaqui but left it out of his final report.) Beyond that, demonstrating Hamilton had misunderstood that original "scrap" of Vattel, he multiplied the evidence for his own position from elsewhere within Vattel's own work.

Hamilton found himself outmaneuvered, and Jefferson won the argument. The treaties stood. Not that it made much difference in the end: America and France ended up fighting a series of high-seas skirmishes just a few years later, and America and Britain fought an all-out war from 1812 to 1815, when the invading redcoats swarmed Washington in April 1814 with torches in hand.

Up from the Ashes

Madison submitted his book list for the Library of Congress in 1783, but it wasn't until the second president of the United States, John Adams, approved the funds that any work on a library for Congress commenced. There was no need at the start. Until 1800 Congress met in Philadelphia, where Benjamin Franklin's Library Company supplied whatever books members needed for research and deliberations. The move to the newly built city of Washington, however, presented a challenge.

Recently cleared from the wilderness, the new capital possessed no urban luxuries such as well-endowed libraries to consult. Recognizing the problem, legislators appropriated $5,000 "for the purchase of such books as may be necessary for the use of congress . . . and for fitting up

THE TAKING OF THE CITY OF WASHINGTON IN AMERICA

a suitable apartment for containing them." President John Adams signed the bill on April 24, 1800, and a committee ordered 740 books (as well as three maps) from an English bookseller. Despite its small beginnings, the library grew apace. By 1812 it boasted more than three thousand books. Then, fourteen years to the day after Adams signed the bill authorizing the library's creation, the British burned it down.

"I learn from the Newspapers that the Vandalism of our enemy has triumphed at Washington over science as well as the Arts, by the destruction of the public library with the noble edifice in which it was deposited," Jefferson wrote his friend, the newspaperman Samuel Harrison Smith, labeling the pyrotechnics "acts of barbarism which do not belong to a civilised age." Of course, anyone could have said something similar. The majority of Americans likely shared Jefferson's outrage. What they didn't share? His singular ability to remedy the situation.

"I presume it will be among the early objects of Congress to recommence their collection," he said to Smith, continuing, "You know my collection, it's condition and extent. I have been 50 years making it, & have spared no pains, opportunity or expence to make it what it is. While residing in Paris I devoted every afternoon I was disengaged, for a summer or two, in examining all the principal bookstores, turning over every book with my own hands, and putting by every thing which related to America, and indeed whatever was rare & valuable in every science."

Beyond his own perusals in Paris, Jefferson had scouts in major European cities on the hunt "for such works relating to America." Regarding books uniquely fitted to American interests, "such a collection was made as probably can never again be effected." Why? "Because it is hardly probable that the same opportunities, the same time, industry, perseverance, and expence, with some knolege of the bibliography of the subject would again happen to be in concurrence."

On top of his collecting overseas of nearly nineteen hundred individual books, as we've seen, he continued to collect when he returned home and, of course, had everything already collected beforehand. Jefferson

Up in smoke: The British set fire to Washington, including the legislative building where Congress housed its library, April 24, 1814. *SOURCE:* LIBRARY OF CONGRESS.

was offering something only he, uniquely in the nation, could offer to his nation, the project of a lifetime: his library.

My earliest memories of Jefferson's gesture were wrong; I'd somewhere picked up the idea that he donated his books to Congress pro bono. No, he sold them to pay off his—and I mentioned this point was coming—crippling debts. As a lump sum, it wouldn't be cheap, but Jefferson offered an unbeatable deal. He figured he had as many as ten thousand volumes in his library. An official inventory came back at much less, just 6,487 volumes. Still, it was the largest personal library in North America. After some internal wrangling and debate from which Jefferson absented himself, Congress picked the collection up for a song: $23,950. It required ten wagons to cart the books to Washington and more than doubled the size of the library destroyed by the British.

"A Common Matrix of Ideas"

I've focused on Thomas Jefferson and James Madison, but similar stories could be told about John Adams, Benjamin Franklin, Alexander Hamilton, or really any of the other major participants in the founding of the country. Even George Washington, considerably less bookish than these, was more bookish than we tend to think.

Adams carried a copy of Cicero in his pocket as a teenager and devoured books on the law, politics, philosophy, and more, collecting a massive library of his own, which he employed exactly as Madison and Jefferson did theirs: When writing, he surrounded himself with books and worked through the piles, finding arguments, establishing precedents, arranging facts, copying quotes, and more.

While the founders and their peers may have fought on any number of issues, they possessed more in common than their differences might suggest. Jefferson, Madison, Adams, Franklin, Washington, and Hamilton, along with Benjamin Rush, Samuel Adams, Thomas Paine, Patrick Henry, George Mason, and the rest, shared patterns of thought shaped by English history and the transatlantic Enlightenment, what Forrest McDonald called "a common matrix of ideas" accessed through "the printed word."

Books written in the classical period, preserved and commented on in the medieval and humanist eras, facilitating the thoughts of philosophers and statesmen into the modern age (some of whom added to the authorial and editorial stream), flowed through the minds and pens of the American founders. An entire system of government and its ongoing operation come out of this tradition, assisted primarily by the technology of books.

Books provide the operating system for institutions and even enable new institutions to be created when old ones fail. As the fire in the capital shows, books are impermanent. But for all their impermanence, they remain among the most permanent of intellectual deposits a culture can make.

Of course, not everyone was included in the wealth. Jefferson lived more than a decade after selling his books. Possessed of "a canine appetite for reading," as he told John Adams in a letter, he kept on collecting—and amassing debts while doing so. He acquired more than nine hundred additional books in that time. All for nought. When he died, they were sold to cover his debts. So were most of the people he enslaved.

Marginalia: Uncommon Impact

The American Revolution was a war of words and ideas. The reason? Motivating people to act—especially those both outgunned and troubled by doubts concerning the radical cause—required justifications, rationales, and arguments. Persuasion had to overcome intuitive, legitimate, and widely shared objections. But more than that, it needed to stoke and stir people's indignation.

Such persuasion often came through speeches and orations, through newspapers, dispatches, and rumors. But in the days leading up to independence, as historian Bernard Bailyn notes, persuasion was most powerfully conveyed through little books. Printed cheaply and often left unbound and uncovered, these pamphlets ranged from five thousand to twenty-five thousand words in length. Printers produced them quickly and easily, and both the writers (who churned them out in torrents) and the market (which craved affordable, digestible content) found the format perfectly suited to both slake and stimulate the public's thirst for pressing concerns.

The majority of these pamphlets were written by community leaders. "The aim of almost every . . . notable pamphlet of the Revolution—pamphlets written by lawyers, ministers, merchants, and planters—was to probe difficult, urgent, and controversial questions and make appropriate recommendations," says Bailyn. We're talking about the sort we already encountered in the chapter above, penned by men like Jefferson, Madison, and Adams. *Common Sense* by Thomas Paine was, however, different on both scores.

Paine, a divorced Methodist lay preacher turned atheist who washed out of most professions he tried (including making corsets, selling groceries, hawking tobacco, and collecting excise taxes), blew into Philadelphia in the fall of 1774 with nothing to waft his sails but a letter of recommendation from Benjamin Franklin. The patriot leader's name meant a lot, as did Paine's sharp and restless mind. Within months of arrival, Paine received an invitation from bookstore owner Robert Aitken to edit his pet project, *Pennsylvania Magazine*. Paine threw himself into the work—and more besides.

Relations with England quickly deteriorated following the Battle of Lexington and Concord in April 1775. Paine began scribbling. By December the draft was done, and Paine's early readers, such as patriot leaders Franklin, Benjamin Rush, and Samuel Adams, approved. Rush suggested the title *Common Sense*, and the first anonymous copies began selling a month later in January 1776.

Few were prepared for the rhetorical fireworks. The explosive prose ignited everyone's curiosity. *Who had written this thing?* Some examples of Paine's art:

- "Society is produced by our wants, and government by our wickedness."
- "Government, even in its best state, is but a necessary evil; in its worst state, an intolerable one."
- "Of more worth is one honest man to society and in the sight of God, than all the crowned ruffians that ever lived."
- "One of the strongest natural proofs of the folly of hereditary right in kings, is that nature disapproves it, otherwise she would not so frequently turn it into ridicule by giving mankind an *ass for a lion.*"

Beyond damning, Paine could also inspire. When readers' eyes fell on such lines as "We have it in our power to begin the world over again" and "The birthday of a new world is at hand," their minds whirred and turned with the possibilities. Of course, not all that whirring and turning was positive, and the possibilities filled some readers with scorn and dread.

"One of the vilest things that ever was published to the world": That's how Tory Nicholas Cresswell characterized *Common Sense*. "Full of false representations, lies, calumny, and treason," the book was calculated "to subvert all Kingly Governments and erect an Independent Republic."

Undersecretary for the colonies and diarist Ambrose Serle read a copy in July. Displeased, he deemed it "a most flagitious Performance, replete with Sophistry, Impudence & Falshood." Serle falsely assumed it had been penned by John Adams, but, regardless of who wrote it, *Common Sense* was "unhappily calculated to work upon the Fury of the Times, and to induce the full avowal of the Spirit of Independence in the warm & inconsiderate"—that is, the hotheaded and unthinking.

Critics had a point. Unlike Jefferson, Madison, or Adams in their work, Paine did not carefully and closely argue the lofty philosophical or gritty legal questions at play in the worsening conflict with England; years later, Adams actually derided the book as "a poor, ignorant, malicious, short-sighted crapulous mass." But Paine was doing something different from these and other men. "Paine's aim," says Bailyn, "was to tear the world apart." *Common Sense* "was meant to overwhelm and destroy."

How so? *Common Sense* spoke to the subterranean fears and hesitations holding back American resistance to the British Crown and then flipped them in ways that subtler and more sophisticated forms of persuasion and argument could not. If we think of the idea grid from earlier, the population tends to live day to day in the *vague and private* quadrant, their hopes and fears inchoate and unexpressed. Paine gave shape to these feelings and put them on paper, channeling some of those impulses toward revolution and giving people permission to discount the rest. By means of pen and print, he dragged those anxieties and resentments into the *precise and public* quadrant, where they could be marshaled for rebellion.

Tory leaders knew he was playing with fire. So did some patriot leaders. But the vast majority of readers cared nothing for their worries. Readers thrilled to Paine's brazen prose and his delegitimization of British rule.

Boosters, including Paine himself, inflated the print success of *Common Sense* from the start. While popular estimates range from 125,000 to 500,000 copies printed, historian Trish Loughran pegs 75,000 as the likely upper limit. Then again, if we factor in the pass-along value, with some of those copies moving through multiple hands, not to mention animating countless coffeehouse and parlor conversations, the pamphlet's reach is still unfathomably long. "Everyone read it," says historian Jill Lepore. As Paine said in a later print run, acknowledging the public's enthusiastic reception of his antigovernment message, "Men read by way of revenge."

Can we credit Paine and his little book with sparking the Revolution? No, but we can certainly credit him with capturing a mood that would justify such an action and expressing it in language that would resonate across the social spectrum. That is to say, *Common Sense* didn't cause the Revolution, but Paine did stamp his indelible mark on it.

Chapter 13

LITERATURE FOR LIBERATION

The Great Book Rescue

James W. C. Pennington watched as the flood waters rose. After a heavy season of snow, a stretch of unseasonably warm temperatures in January 1841 caused the ice to melt. Rain fell, and the Connecticut River began flooding the adjacent city of Hartford, where he pastored a church and taught school.

Luckily for him, Pennington and his wife lived in a tiny apartment on the west side of town, away from the flooding and out of harm's way. Unluckily for him, he kept a small study on the third floor of a large home on Front Street, right next to the river. Up those three flights of stairs Pennington not only stored his personal library but also kept a book he was writing. The completed manuscript, ready to be typeset, printed, and published, now risked being swept away by the currents.

At first unaware of the danger, Pennington walked to his library only to be blocked by the flood. He'd need a boat to reach his books. By then the other tenants of the building had either scrambled out or moved up to the second story. All along the waterfront the first stories of homes were submerged, and people awaited rowboat rescues from their second- and third-story windows.

In one of those boats rowed a couple of Pennington's parishioners. Hearing the plight of their pastor, they paddled their way to the house, breached a window, and rescued both his library and his precious manuscript—just in the nick of time, as it happened. The river didn't crest for another day, rising all the while and sweeping houses off their

foundations. More than a hundred Hartford families lost homes in the flood. But Pennington's manuscript survived, as did the local printer's shop, and a month after the deluge, his book was published.

At some level, every book is unique, making each's appearance in the world a novelty worthy of special notice. That was certainly true of Pennington's project—in fact, doubly so. *A Text Book of the Origin and History of the Colored People* became the first book of African American history published in the country. What's more, it was written by a fugitive slave whose education was illegal in the state of his birth.

The Talking Book

Their form, function, and use are so familiar that we never think twice about how books work. But consider someone with little prior experience with the device. By what mysterious mechanisms does this peculiar machine operate?

James Albert Ukawsaw Gronniosaw wanted the book to speak to him, as he saw it do with others. Born in West Africa in the first part of the eighteenth century and sold into slavery as a young man, Gronniosaw found himself on a Dutch slave ship headed to Barbados. The ship's captain read prayers to the crew on Sundays. "When first I saw him read," said Gronniosaw in his 1772 autobiography, one of the first examples of the slave narrative genre, "I was never so surprised in my whole life as when I saw the book talk to my master; for I thought it did, as I observed him to look upon it, and move his lips. I wished it would do so to me."

When the captain had finished reading, he put the book away, and Gronniosaw sneaked behind him to retrieve it. "I follow'd him to the place where he put the book," he said, "being mightily delighted with it, and when nobody saw me, I open'd it, and put my ear down close upon it, in great hope that it would say something to me; but [I] was very sorry and greatly disappointed when I found it would not speak."

Others shared the assumption that the book somehow talked to its beholder; the first several slave narratives on record refer to this notion.

Reverend J. W. C. Pennington. *SOURCE:* SCHOMBURG CENTER FOR RESEARCH IN BLACK CULTURE, NEW YORK PUBLIC LIBRARY.

Olaudah Equiano's 1789 autobiography takes readers aboard a ship from Virginia bound for England, where he witnessed his enslaver and a friend passing time with a book. "I had a great curiosity to talk to the books, as I thought they did," Equiano confessed. "I have often taken up a book, and have talked to it, and then put my ears to it when alone, in hopes it would answer me; and I have been very much concerned when I found it remained silent."

It's possible that such statements of what Henry Louis Gates Jr. calls "the trope of the talking book" represent a form of put-on. But whether confessed for real or concocted for effect, the naivety is understandable. It takes me back to Clarke's Third Law: "Any sufficiently advanced technology is indistinguishable from magic." A book communicates—but how, exactly? Without knowledge of the means by which the strokes and squiggles on the page cue associated ideas and sounds in the mind and mouth, the communicative powers of a book operate behind a shroud.

Gronniosaw was eventually sold to a Dutch minister in New York named Theodorus Jacobus Frelinghuysen (Freelandhouse in the autobiography), a figure in the First Great Awakening. He and his wife sent Gronniosaw to school to unravel the mystery of the talking book. Though Gronniosaw was at first "uneasy" about going, "I came to like it," he said, "and learnt to read pretty well." Gronniosaw wasn't alone. Anglicans sponsored a school for New York's enslaved; in 1707, more than a hundred pupils attended. Another Anglican effort in New York saw upward of seventy educated at Trinity Church between 1733 and 1747.

Still, the mechanics of the talking book remained out of reach for most, and literacy among the enslaved would become increasingly contested and opposed by White enslavers and their allies in government. Before long, the idea of enslavers sending the enslaved to school to learn to read and write would be every bit as unthinkable as the enslaved found the promise of literacy irresistible, especially in the South.

And for the very same reason: The enslaved could, would, and did put reading and writing to work for their own purposes—purposes at odds with those of the people who claimed to own them. The problem—or, more accurately, the promise—revealed itself early on. Much of literacy's potential for African Americans was telegraphed in the brief but brilliant life of Phillis Wheatley.

Free Verse

Kidnapped as a child from her home in West Africa and renamed for the slave ship she disembarked from in Boston at seven or eight years of age in 1761, Phillis Wheatley demonstrated a preternatural gift with language. There was, for instance, the time her owners noticed the young girl "endeavoring to make letters upon the wall with a piece of chalk or charcoal." Helped by her mistress Susanna and Susanna's daughter Mary, Wheatley quickly learned both to read and to write, and not just workaday stuff.

In 1767, one of her poems first found its way into print. It commemorated the survival of two men nearly drowned in a terrible storm at sea. The opening verses:

> *Did Fear and Danger so perplex your Mind,*
> *As made you fearful of the whistling Wind?*
> *Was it not Boreas knit his angry Brow*
> *Against you? or did Consideration bow?*
> *To lend you Aid, did not his Winds combine?*
> *To stop your Passage with a churlish Line,*
> *Did haughty Eolus with Contempt look down*
> *With Aspect windy, and a study'd Frown?*

Over the next two decades, she wrote as many as 145 poems. The first to win her real fame came upon the 1770 death of George Whitefield, the famous evangelist and preacher of the movement known as the Great Awakening. Wheatley elegized the minister with forty-seven lines of verse. A sample:

> *Thou didst in strains of eloquence refin'd*
> > *Inflame the heart, and captivate the mind.*
> *Unhappy we the setting sun deplore,*
> > *So glorious once, but ah! it shines no more.*

The poem won Wheatley fans on both sides of the Atlantic, ultimately connecting her to the patron who would help her publish a celebrated collection of her work a year later in 1771, *Poems on Various*

Subjects, Religious and Moral. With the book's publication, Wheatley became the first African American woman published in North America and among the continent's most successful poets. But her book also occasioned a crisis of sorts: As both an enslaved woman and a published author, Wheatley embodied a societal contradiction.

The arguments for keeping Blacks in the condition of subservience depended on their being seen as inferior to Whites. "We are," as Black mathematician and scientist Benjamin Banneker said in a 1791 letter to Thomas Jefferson, "a race of Beings who have long laboured under the abuse and censure of the world . . . considered rather as brutish than human, and Scarcely capable of mental endowments." Of course, Banneker's own life and accomplishments undercut any such assumptions; the same applied to Wheatley. They both stood as living proof to the contrary.

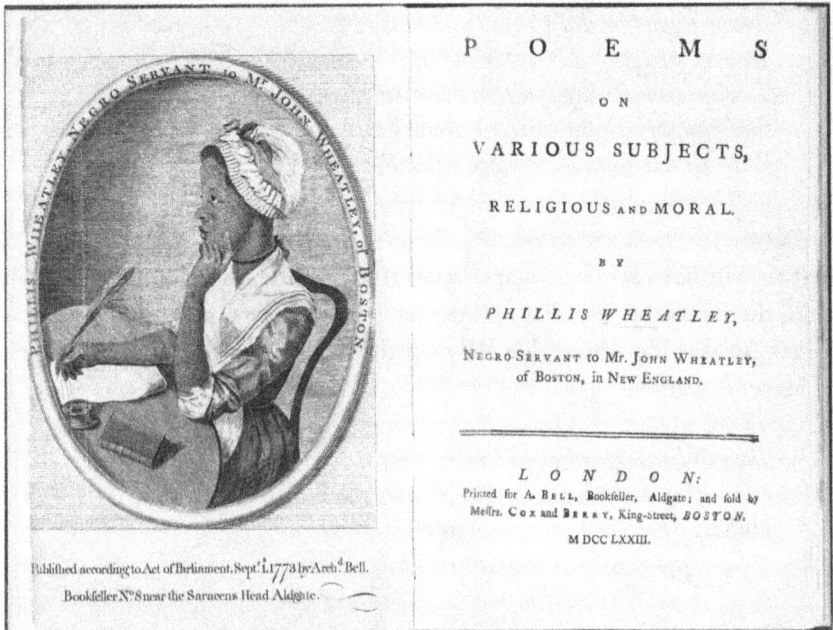

Phillis Wheatley, writing free verse. Engraving by Scipio Moorhead. *SOURCE:* SCIENCE HISTORY IMAGES, ALAMY.

The easiest response to such walking, talking arguments? Denial. Some readers expressed doubts that Wheatley had written her poems, though her skills were easily confirmed. Eighteen of Boston's leading citizens, including the royal governor and lieutenant governor, seven ministers, even the merchant John Hancock—he of the extravagant, declaratory signature—validated her abilities and attached their names as testimony to her book. "She has been examined by some of the best Judges, and is thought qualified to write them," announces a prefatory note to the reading public in *Poems on Various Subjects*.

While validating Wheatley, that same notice invalidated the country's peculiar institution. If she really were an "uncultivated barbarian from Africa" and "a slave in a family in this town," how did she produce line after line of elegant verse? The only way to preserve the foundational lie on which race slavery was built was to downplay Wheatley's accomplishments.

In a passage that now tarnishes his reputation, Thomas Jefferson denied poetic abilities to African Americans as a race. "Never yet could I find that a black had uttered a thought above the level of plain narration," he wrote in *Notes on the State of Virginia*, continuing, "Misery is often the parent of the most affecting touches in poetry. Among the blacks is misery enough, God knows, but no poetry. . . . Their love is ardent, but it kindles the senses only, not the imagination. Religion indeed has produced a Phyllis Whately [*sic*]; but it could not produce a poet. The compositions published under her name are below the dignity of criticism."

Jefferson dismisses Wheatley's work as unpoetic, mere religious twaddle unworthy of critical engagement. But he'd have to say that, right? The entire project of slaveholding Virginia—and America as well—assumed a deficit of African humanity. Wheatley's demonstrable talent and abilities rebutted the argument. By denying her art, he keeps her a brute.

What of Roman slaves? Jefferson admits "their slaves were often their rarest artists." What's more, these slaves were so well educated they were retained to teach the children of their owners. "Epictetus, Diogenes, Phaedon, Terence, and Phaedrus, were slaves," admits Jefferson. "But they were of the race of whites. It is not their condition then, but nature, which has produced the distinction." It's as if he's responding directly to the prefatory note in *Poems on Various Subjects*, which mentions that

Wheatley had developed her gifts while "under the disadvantage of serving as a slave."

If one could grant her accomplishments, the fact of her enslavement would serve only to amplify them. But Jefferson can deny any attainment by saying Wheatley was constitutionally incapable of the feat. Of course, he must also overlook the fact that Terence was Black, a fact to which Wheatley appealed, along with the many classical allusions evident from the very first poem in the collection. Should readers dignify the poems enough to read them critically, they might find plenty to contradict Jefferson's dismissal.

And so he leaves himself an out every bit as amusing as it is petty: While denying Wheatley has written anything worth reading, Jefferson suggests she hasn't written anything at all. The phrase "published under her name" allows him to say the poems, if they are worth reading, aren't hers anyway.

Why go to such lengths to discredit a single poet? "Wheatley takes up only a few lines at the center of Jefferson's lengthy defense of the Americans at their weak point—as enslavers—but in a sense it is all about her," says her biographer David Waldstreicher, "made necessary by her fame." Wheatley refutes chattel slavery's most precious myth—namely, that those in chains deserved their bonds through their very nature. As one college president asked in 1810, "What planter can write so well?"

The contradictions demanded resolution. How could Wheatley's work reflect such genius while her natural state was supposedly so inferior? With her publication, she owned a copyright, but how could she claim ownership while also being owned? There were too many internal conflicts within Wheatley's condition, and her owners knew it. They chose to emancipate her—or maybe we should say she won her freedom—in 1778.

Literacy forced a reckoning and would continue to do so. In response, some states, particularly those in the South, banned instruction in reading and writing for the enslaved.

Crackdown

Born in 1796 or 1797 in Wilmington, North Carolina, David Walker escaped slavery through his mother's legs. Enslavers claimed all children born to female slaves. Walker's father was enslaved, but his mother was free; consequently, so was he. We know relatively little about his upbringing, but he traveled extensively, observing the condition of fellow Blacks and wishing they could share his free status. We also know he learned to read and write and leveraged those skills to help realize his desire after migrating to Boston around 1825.

Moving in abolitionist circles, he wrote and sold subscriptions for *Freedom's Journal*, the first Black-owned newspaper in the United States. Published in New York and distributed in nearly a dozen states, Washington, DC, Canada, Haiti, and Europe, the inaugural edition declared, "We wish to plead our own cause. Too long have others spoken for us." The paper folded after a brief run, but that didn't slow down Walker. In 1829, he penned a widely distributed treatise so incendiary it could singe the paper it was printed on.

Walker wrote *Appeal to the Coloured Citizens of the World* to inspire opposition to slavery. He argued that slavery degraded Black people and kept them in bondage through enforced ignorance. Since keeping the enslaved uneducated was essential to maintaining this state of oppression, literacy was seen as a threat to enslavers. "For colored people to acquire learning in this country," said Walker,

> *makes tyrants quake and tremble on their sandy foundation. Why, what is the matter? Why, they know that their infernal deeds of cruelty will be made known to the world. Do you suppose one man of good sense and learning would submit himself, his father, mother, wife and children, to be slaves to a wretched man like himself, who, instead of compensating him for his labours, chains, hand-cuffs and beats him and family almost to death, leaving life enough in them, however, to work for, and call him master? No! no! he would cut his devilish throat from ear to ear, and well do slave-holders know it. The bare name of educating the coloured people, scares our cruel oppressors almost to death.*

Walker published his treatise in Boston in September. Within months it was all over, rapidly finding those "Coloured Citizens" to whom it was addressed. Police in Savannah, Georgia, seized sixty copies of the tract after an uneasy African American minister encountered a copy and gave it to officials; the First, Fourth, and Fifth Amendments to the US Constitution had little purchase in the states at the time and even less when the Georgia legislature immediately passed a bill "to prevent the circulation of written or printed papers within this State calculated to excite disaffection among the coloured people of this state, and to prevent said people from being taught to read or write." What counted as "disaffection"? Legislators wrote the law broadly enough to include actions ranging from inciting insurrection to getting uppity. Punishment? Death.

Within weeks, Louisiana joined Georgia to ban not only circulating literature unfriendly to slavery but also teaching the enslaved to read or write. Officials needed to stanch the flow of subversive ideas. Reading and writing represented a tool the enslaved could use to assert their humanity and freedom—even with violence if circumstances arose. For his part, Walker said he hoped White slaveholders would repent but figured they wouldn't; in a biblical image of divine recompense, he said they'd already "filled the cup" against themselves and would only dig in. He was right.

(A note in passing: Walker's *Appeal* is full of both implicit and explicit biblical and theological language. His use of "sandy foundation" is, for instance, a biblical metaphor taken from one of Christ's parables. Some of his strongest language is leveled against White Christians whose actions betray the tenets of their mutually held faith.)

Of course, preventing discontent among the enslaved by banning Walker's book was like holding back the tide. Slaves had access to anti-slavery messages all over: not only the words of Phillis Wheatley, the implicit spirit of whose work caused Jefferson distress, but also explicit abolitionist literature already infiltrating the South. Beyond that, there was the Bible itself.

Vast numbers of Blacks in America identified as Christian—some, such as the enslaved Congolese who ended up near Charleston, South Carolina, going all the way back to Africa. Sixty of these slaves near the Stono River outside Charleston revolted in 1739, seemingly inspired by both their Catholic faith and Spanish propaganda that they, as literate in Portuguese, managed to read. The uprising failed, but twenty-three slave owners died before it was quashed. Believing the slaves had used writing to coordinate their activities, the following year the colonial legislature banned anyone teaching slaves to write or, if they already knew how, employing them to do so.

While South Carolina's 1740 Slave Code outlawed instruction in writing, it allowed individual masters to teach reading if they chose. But as reading precipitated further crises, such discretion proved too risky. In 1822, for instance, the freedman Denmark Vesey masterminded a slave revolt not far from the 1739 Stono Rebellion, directly inspired by his Bible reading. The Exodus formed a central theme. Casting enslavers as Egyptians and the enslaved as Israelites, he stoked the fires of liberation. The violent insurrection failed to materialize; it was stopped when authorities caught wind of the plot after insiders revealed the plan. Vesey and nearly forty conspirators were hanged.

Nat Turner's 1831 revolt up the coast in Southampton County, Virginia, cinched the knot. Lasting two days and resulting in the deaths of at least fifty-five Whites, the bloody uprising was inspired in part by Turner's reading of the Bible. When captured, Turner was armed with both a sword and his copy of the scripture. In themes as well as verses to describe his role as a leader and the tactics employed by him and his followers.

In the wake of Turner's rebellion, Whites grew outright fearful of slave literacy. What was going on in those very human minds, behind those very brown eyes? Harriet Jacobs, a slave whose mistress taught her how to read, found her home searched by a ragtag local militia: unlettered "low whites" with "no negroes of their own to scourge." They hunted for letters, papers, and books, hoping to find any evidence of literacy, their captain waiting in the wings.

"We's got 'em! We's got 'em!" the searchers exclaimed when they finally turned up something incriminating. "Dis 'ere yaller gal's got letters!" Another plot? The illiterate men couldn't read for themselves and passed the letters to their captain to decode the script. But, no, the letters contained poetry from one of Jacobs's friends. "When their captain informed them of their contents, they seemed much disappointed," said

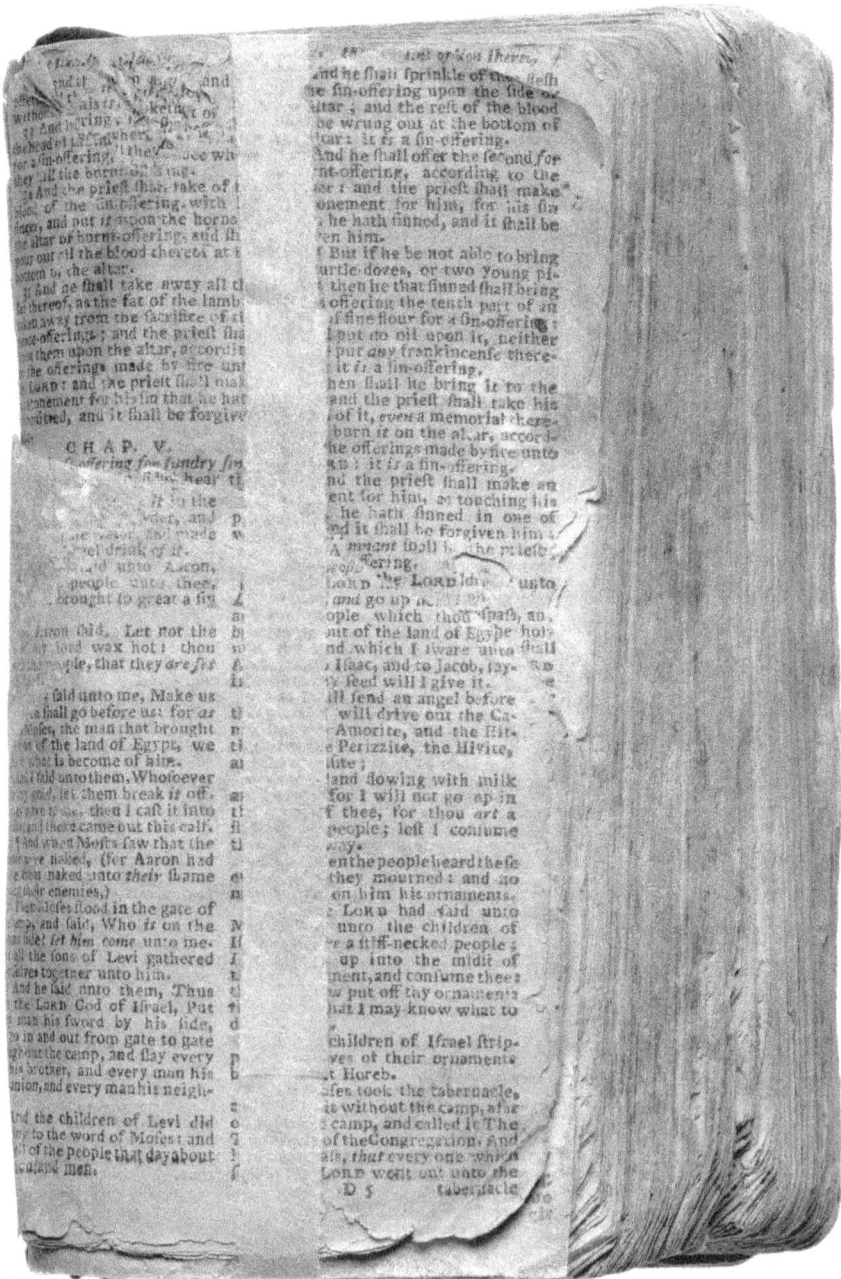

Nat Turner's well-used Bible, missing the front and back covers, the first two books of Moses, and Revelation. *SOURCE:* COLLECTION OF THE SMITHSONIAN NATIONAL MUSEUM OF AFRICAN AMERICAN HISTORY AND CULTURE.

Jacobs. Then again, if her friends wrote her poetry, that proved Jacobs could read, no?

"Can you read them?" asked the captain. Jacobs answered yes, and the man "swore, and raved, and tore the paper into bits."

Between domestic agitation caused by abolitionist literature and the dangers posed by slaves whose eyes might land on inciting prose—even the Bible—a wave of laws targeting both abolitionist literature and literacy itself passed in Southern states. "Laws have been recently passed in all these States making it penal to teach slaves to read," said South Carolina Governor James Henry Hammond in an 1845 open letter to English abolitionist Thomas Clarkson, adding,

> *Do you know what occasioned their passage, and renders their stringent enforcement necessary. I can tell you: it was the abolition agitation. If the slave is not allowed to read his Bible, the sin rests upon the Abolitionists; for they stand prepared to furnish him with a key to it, which would make it, not a book of hope and love and peace, but of despair, hatred and blood; which would convert the reader, not into a Christian but a Demon. To preserve him from such a horrid destiny, it is a sacred duty which we owe to slaves, not less than to ourselves, to interpose the most decisive means. . . . Allow our slaves to read your pamphlets, stimulating them to cut our throats! Can you believe us to be such unspeakable fools.*

Literacy also enabled much subtler, private rebellions—such as escape. An enslaved man in Maryland, John Thompson, came across an oration of John Quincy Adams. "While reading this speech, my heart leaped with joy," he reported. Thompson hid his copy and repeatedly sneaked away on Sundays to read it alone in the woods. "I . . . read it until it was so worn that I could scarce make out the letters," he said. "I then found out that there was a place where the negro was regarded as a man, and not as a brute; where he might enjoy the 'inalienable right of life, liberty, and the pursuit of happiness'; and where he could walk unfettered throughout the length and breadth of the land." Adams's words inspired Thompson's flight north. Literate slaves could forge passes for themselves and others to help aid their escapes; many did.

Newspaper notices for runaways provide evidence of literacy among enslaved populations. In these notices, slaveholders identified characteristics of runaways to aid in their return, including whether they could read, write, or both. An analysis of these notices reveals that 7 percent of Virginia runaways were literate, as were 20 percent of runaways from one area of Kentucky. The numbers are considerably lower when averaged across the South, just 2–3 percent, though numbers drawn from notices likely undercount the true totals; masters were sometimes the last to discover their people knew how to read and write. Some slaves could read but "ke' dat up deir sleeve," one former slave admitted; "dey played dumb lack de couldn't read a bit."

But how did they learn?

Fugitive Students

Gronniosaw, Wheatley, and Jacobs all picked up the skill when their masters and mistresses taught them, a "privilege, which so rarely falls to the lot of a slave," as Jacobs said. But it was more common than we might guess, despite the mounting legal and social obstacles in the early nineteenth century.

One study of testimonies left by 272 literate former slaves reveals that owners often initiated their training, most typically for religious instruction. Why? Some owners hoped to encourage docility through Christian teaching, though as evidenced by David Walker, Denmark Vesey, and Nat Turner—and feared by Governor James Henry Hammond—the Bible could be a double-edged sword; slaves found their plight mirrored in that of the enslaved Israelites and eagerly awaited similar deliverance. Still, many slaves were forced to learn on the sly, becoming "fugitive learners," as Harvard professor Jarvis R. Givens refers to them.

Susie King Taylor was, as she said, "born under the slave law in Georgia in 1848." Her grandmother raised her and saw to her education, along with that of her brother. Each day the kids would take their books, "wrapped in paper to prevent the police or white persons from seeing them," to visit a widowed free Black woman in their neighborhood. Their

Susie King Taylor, fugitive learner. *SOURCE*: PIEMAGS/CMB, ALAMY.

Susie King Taylor.

secrecy went beyond hiding contraband books. As the children arrived at the woman's home, they went around back to the kitchen one at a time to avoid arousing the suspicions of White neighbors. Helped by her daughter, the free woman taught twenty-five to thirty kids at a time in her clandestine schoolhouse.

Fugitive students like Taylor met in each other's homes, in groves and glades, under the cover of night, and even in holes in the ground to learn their letters and how their artful arrangement could open up worlds beyond the control of White society.

Slaves often found themselves under lax supervision on Sundays. Informal schools emerged to make use of the time, and a minister or another person in the community would teach the unlettered. As a young man, while still enslaved, Frederick Douglass helped out in such a Sabbath school. "A pious young man named Wilson," described by Douglass as White in one of his three memoirs, secured the use of a free Black man's home, "mustered up a dozen old spelling-books and a few Testaments, and we commenced operations, with some twenty pupils in our school." Unfortunately, the school closed almost as soon as it started. "I learned there were some objections to the existence of our school," he said. On the second or third Sunday a mob "armed with sticks and other missiles" broke up the gathering. "One of this pious crew told me that as for me, I wanted to be another Nat. Turner," Douglass recalled, "and that, if I did not look out I should get as many balls in me as Nat. did into him." *Balls*, as in lead, as in rifle shot: The men threatened to kill Douglass if he kept teaching.

Educational disruption was the norm, and threats such as these were backed up by brutal insistence. "Slaves . . . caught writing or learning to write," said historian Allen Dwight Callahan, "might have their fingers cut off." Such threats were ubiquitous; after reading thousands of slave narratives, Janet Cornelius found testimony of amputation in Georgia, Mississippi, South Carolina, and Texas. "The first time you was caught trying to read or write," said one slave from Madison County, Georgia, "you was whipped with a cow-hide, the next time with a cat-o-nine-tails and the third time they cut the first jint offen your forefinger." The uncle of another slave stole a book to study; when discovered, his master "had the white doctor take off my Uncle's fo'finger right down to de 'fust jint." Another slave recalled his father being whipped to death for teaching other slaves to read.

Regardless, countless individuals among the enslaved deemed the promise of liberation through literacy worth the risk. Two slaves on one plantation learned from itinerant workers and practiced by scrawling on the barn walls—which their master saw. Undeterred, they kept at it. One of the two got skilled enough to write letters and try sending them through the mail. He was whipped for the attempt but later forged a pass and escaped.

Many slaves picked up reading and writing as kids from literate White children. Neighborhood kids taught William Anderson his letters, which he practiced in the sand. He somehow bought a book and hid it in his hat. Sometimes, he said, he stashed it in "the leaves or earth for fear of the lash." In the nursery, hidden in a hayloft, tagging along with schoolkids—enslaved children found all sorts of contexts for picking up know-how from their playmates and their masters' families. One boy swapped apples for reading lessons.

As a child, Frederick Douglass worked a similar trade, only he swapped bread. "I used to carry, almost constantly, a copy of Webster's spelling book in my pocket," he said. Whenever he ran an errand or had a moment of playtime, he met friends on the street, pulled them aside, and asked for a lesson. "For a single biscuit, any of my hungry little comrades would give me a lesson," he said, though it's worth mentioning that some did it for free—never underestimate the thrill of doing something naughty, even if it's only illegally teaching your Black friend how to find his way down a page on his own.

It was his owners' fault, really. Douglass first learned his alphabet and the rudiments of reading from his naive mistress, Sophia Auld. She was soon warned that literacy would "spoil" him as a slave and stopped her lessons. Too late. Douglass had the taste, and once he learned, he couldn't be stopped.

Douglass had to read on the sly but kept up the practice whenever he could. Eventually, he found and purchased a copy of Caleb Bingham's *The Columbian Orator*, a primer on grammar and rhetoric that contained, along with speeches by Cicero, William Pitt, and George Washington, two passages that transformed his life. "I found in it," he said,

> *a dialogue between a master and his slave. The slave was represented as having run away from his master three times. The dialogue represented the conversation which took place between them, when the slave was retaken the third time. In this dialogue, the whole argument on behalf of slavery was brought forward by the master, all of which was disposed of by the slave. The slave was made to say some very smart as well as impressive things in reply to his master—things which had the desired though unexpected effect; for the conversation resulted in the voluntary emancipation of the slave on the part of the master.*

Imagine reading that as an enslaved person—all the arguments for and against your predicament presented for you to engage, internalize, and then externalize as opportunities arose.

The second passage? "In the same book, I met with one of Sheridan's mighty speeches on and in behalf of Catholic emancipation," he said. While the *Orator* featured a selection from Anglo-Irish statesman and playwright Richard Brinsley Sheridan, Douglass seems to have conflated it with another speech by another Irishman, Arthur O'Connor, also in the *Orator*: "Part of Mr. O'Connor's Speech in the Irish House of Commons, in Favor of the Bill Emancipating the Catholics." But the mix-up is immaterial to the impact of Douglass's reading "Dialogue between a Master and Slave" and O'Connor's speech:

> *These were choice documents to me. I read them over and over again with unabated interest. They gave tongue to interesting thoughts of my own soul, which had frequently flashed through my mind, and died away for want of utterance. The moral which I gained from the dialogue was the power of truth over the conscience of even a slaveholder. What I got from Sheridan was a bold denunciation of slavery, and a powerful vindication of human rights. The reading of these documents enabled me to utter my thoughts, and to meet the arguments brought forward to sustain slavery. . . . The more I read, the more I was led to abhor and detest my enslavers.*

Reading gave Douglass the language he required to shape thoughts of his own freedom. He eventually learned to write as well, escaped to the North, and used his skills to become the greatest American orator of the nineteenth century and a champion of human liberty admired to this day.

A Right Reclaimed, an Accusation Leveled

When the escaped slave James Pennington fled his captivity at nineteen years old, he knew next to nothing of reading and writing. He could scrawl a few letters, and very neatly, but that was all. Luckily for him, not only had his first hosts in the North, the Quakers William and Phebe Wright, helped hundreds of fugitive slaves make their way to freedom, but William was also a former schoolteacher and set about filling in the gaps. "We can soon get thee in the way," he assured.

Pennington learned the basics: reading, writing, arithmetic. William Wright also taught him astronomy. And he inspired his student with the stories of famous former slaves who had achieved greatness through learning, including two we've already met: Phillis Wheatley and Benjamin Banneker. Pennington rallied under their example and tried teaching himself public speaking, sensing the importance for his future.

He soon made his way to New York and settled in Brooklyn. Sabbath schools and night schools fed his appetite for learning. He quickly outstripped the available curriculum and began studying rhetoric, logic, Latin, and New Testament Greek—on his own without a tutor, from books, naturally. (Take that, Socrates.)

He eventually became a teacher and preacher, studied at Yale, and served as a leader in the Colored Convention Movement, arguing, among other things, that Blacks should not return to Africa but should stay in their country, America, and claim the freedom that was rightfully theirs by birth and toil. Then there were the books, essays, and speeches—some of which were published by none other than Frederick Douglass in the newspaper he edited. The two men befriended each other when Douglass moved to New York; as a minister, Pennington married Douglass and his wife, Anna Murray.

Frederick Douglass, working for the cause. *SOURCE*: SCIENCE HISTORY IMAGES, ALAMY.

Despite all these accomplishments, when Pennington finally sat down to write his memoir, he included a specific and un-addressable complaint. "There is one sin that slavery committed against me, which I never can forgive," he said, adding,

> It robbed me of my education; the injury is irreparable; I feel the embarrassment more seriously now than I ever did before. It cost me two years' hard labour, after I fled, to unshackle my mind; it was three years before I had purged my language of slavery's idioms; it was four years before I had thrown off the crouching aspect of slavery; and now the evil that besets me is a great lack of that general information, the foundation of which is most effectually laid in that part of life which I served as a slave. When I consider how much now, more than ever, depends upon sound and thorough education among coloured men, I am grievously overwhelmed with a sense of my deficiency, and more

especially as I can never hope now to make it up. . . . I shall have to go to my last account with this charge against the system of slavery, "Vile monster! thou hast hindered my usefulness, by robbing me of my early education."

Against the backdrop of this complaint, it's astonishing to contemplate all Pennington did accomplish. But despite his achievements, think of all the potential that was taken, all the promise irretrievably lost, all the possibilities left unfulfilled. Then multiply that crime by millions of minds similarly deprived.

Marginalia: Malcolm X's Alma Mater

When asked by a British writer about his alma mater, Malcolm X answered, "Books." The response would have surprised anyone who knew him growing up—back when he was still known as Malcolm Little, before he joined the Nation of Islam, before he went to jail.

Born in Omaha, Nebraska, in 1925, Malcolm spent his boyhood in and around Lansing, Michigan. Earl Little, a Baptist preacher and evangelist for Marcus Garvey's message of Black empowerment, was a stern and insistent father. Always agitating for his views, he was killed by White supremacists when Malcolm was just a child; the official story was that a streetcar had run him over, but the family never believed it.

The Littles were already in difficult straits but had managed to stay strong. Earl's murder pushed them over the edge. Eventually the kids were fostered out, and mom Louise was committed to a mental hospital, where she stayed for the next twenty-six years.

Between the murder and the collapse, Malcolm's older brother Wilfred began reading. "His head was forever in some book," Malcolm recalled. But Malcolm was disinclined to study. His eighth-grade English teacher encouraged White students to think big about their futures. Not Malcolm, who enthused about practicing law. "A lawyer," his teacher objected, "that's no realistic goal for a nigger." He suggested carpentry instead.

"It was then I began to change, inside," said Malcolm. He quit school and moved to Boston, later settling in Harlem, where the would-be lawyer expressed his disillusionment with the system by breaking the law he previously imagined serving. Though he worked legitimate jobs, he gravitated toward crime: drugs, prostitution, theft, and even armed robbery. He was eventually caught. Before turning twenty-one, Malcolm began a ten-year prison sentence. It was the best thing that ever happened to him.

By the time he entered prison, Malcolm had forgotten much of what he'd learned in school. "I didn't know a verb from a house," he said. That changed when a fellow inmate suggested he try some correspondence courses and make use of the prison library. Malcolm's bookish brother Wilfred also played a part. "While you're in there," he advised Malcolm, "spend time in that library." Malcolm started reading, took an English course, and even began Latin lessons.

To expand his vocabulary, he began hand-copying the dictionary, an idea he seemed to have picked up from his mother. "As my word-base broadened," Malcolm recalled, "I could for the first time pick up a book and read and now begin to understand what the book was saying. Anyone who has read a great deal can imagine the new world that opened." Now Malcolm's face, just like brother's, was forever in a book.

He read as much as fifteen hours a day: history, philosophy, science, mythology, and more. He read, among many others, W. E. B. Du Bois, Will Durant, Immanuel Kant, Gregor Mendel, John Milton, Friedrich Nietzsche, Baruch Spinoza, Harriet Beecher Stowe, H. G. Wells, Carter G. Woodson, even Dale Carnegie. He called these "prison studies" his "homemade education," and his voracious appetite for literature gave him the sense, sound, and style of someone who stayed in school well past the eighth grade.

"I have often reflected on the new vistas that reading opened to me," he said. "I knew right there in prison that reading had changed forever the course of my life. As I see it today, the ability to read awoke inside me some long dormant craving to be mentally alive."

Through reading, Malcolm had reached beyond what W. E. B. Du Bois would call the Veil. "I sit with Shakespeare and he winces not," Du Bois says in a passage of *The Souls of Black Folk* reminiscent of Niccolò Machiavelli's nighttime encounters with his favorite authors but loaded with an accusation: "Across the color-line I move arm in arm with Balzac and Dumas, where smiling men and welcoming women glide in gilded halls. From out the caves of evening that swing between the strong-limbed earth and the tracery of the stars, I summon Aristotle and Aurelius and what soul I will, and they come all graciously with no scorn nor condescension. So, wed with Truth, I dwell above the Veil."

Malcolm's glimpses above the Veil empowered him to operate within it. Thanks to his reading, he became a fierce advocate for fellow Blacks and a harsh opponent of what he called "the racist malignancy" in White America. He became the most recognizable radical civil rights proponent of the era. Controversial both within and outside the movement, on February 21, 1965, he was assassinated on stage while speaking.

In life, few could match Malcolm X's ferocious literary habits. Still, like James W. C. Pennington, he regretted never finishing school. "You can believe me that if I had had the time right now," he said, "I would not be one bit ashamed to go back into any New York City public school and start where I left off at the ninth grade, and go on through a degree." There's a sadness in those words, a recognition of loss. But it's also true that no schooling in the world would have been ample enough for Malcolm X. He was never done learning.

Chapter 14

SEEING WITH OTHER EYES

"My Children Should Not Read Novels"

After her mother died, little Harriet found solace in books. Though only four at her mother's passing, she took to reading early and easily. "She has been to school all summer," her older sister wrote, "and has learned to read fluently." By then she was five, and Harriet walked hand in hand with her four-year-old brother to the school yard. Her father, a respected preacher, bragged that she was "intelligent & studious." True, she was a little "odd"—his word—but she was also a "great genius."

Good thing her father had plenty of books to choose from. He kept a library in the attic of their Litchfield, Connecticut, home. "This room had to me the air of a refuge and a sanctuary," she recalled. "Its walls were set round from the floor to ceiling with the friendly, quiet faces of books." While friendly, most of Dad's tomes bored Harriet to distraction. "There were Bell's Sermons, Bonnett's Inquiries, Bogue's Essays, Toplady on Predestination, Boston's Fourfold State, Law's Serious Call, and other works of that kind," she said. "These I looked over wistfully, day after day, without even a hope of getting something interesting out of them. The thought that father could read and understand things like these filled me with a vague awe."

Despite the dross, Harriet occasionally found gold. She thrilled at discovering Reverend Cotton Mather's *Magnalia Christi Americana*, first published in England the prior century, detailing episodes of New England's church history. An American publisher released a new two-volume edition in 1820, when Harriet was nine. Mather encouraged his New England readers to treat their slaves well. Harriet might have registered the point, but she was mostly drawn to Mather's stories. "What wonderful stories those!" she said.

While in school, she studied history, philosophy, rhetoric, composition, and more, but stories drew her in and fired her imagination. Rummaging through storage barrels in the attic, she found tracts, pamphlets, and various harangues for moral and religious reform. But what was that? Peeking through these castoffs, she spied unbound pages of *Don Quixote* and a complete copy of *The Arabian Nights*, which she rescued from the pile and devoured.

In time she accumulated a stash of her own books, copies of *The Tempest*, *Ivanhoe*, and others. Her father was ambivalent about fiction but had given her a complete set of Sir Walter Scott's work as part of a bequest. "I have always said that my children should not read novels," he said, "but they must read these." Harriet later became a novelist herself. In one of her novels, she describes "the library of a well-taught young woman of those times." It included issues of Joseph Addison's *The Spectator*, *Paradise Lost*, Shakespeare, *Robinson Crusoe*, and Samuel Richardson's seven-volume novel *Sir Charles Grandison*, plus the Bible and works of Jonathan Edwards.

We can imagine these and other books as Harriet reflected on the joy of hiding away in her father's attic library: "Here I loved to retreat and niche myself down in a quiet corner with my favorite books around me." All of Harriet's reading, perhaps especially the fiction, sharpened her moral sense. She could read and discuss the Bible, sermons, tracts, and treatises, but stories ignited her imagination. And she knew they could fire other people's imaginations as well—far more effectively than all those tracts and sermons.

So, as an adult, when she decided Mather didn't go far enough and she wanted to see slavery abolished, not merely softened, Harriet Beecher Stowe wrote *Uncle Tom's Cabin*. For someone whose father mostly disapproved of fiction, Stowe penned one of the most powerful and consequential novels ever written.

Harriet Beecher Stowe with a copy of *Uncle Tom's Cabin*. Lithograph by Marie Alexandre Alophe. SOURCE: RIJKSMUSEUM, WIKIMEDIA COMMONS.

Mrs BEECHER STOWE

Auteur de LA CASE DU PÈRE TOM.

Maison GOUPIL & C.ie
Paris . Londres.

Paris, Imp. Impériale Goupil & C.ie   New-York, Published Goupil & C.o

Stowe began writing *Uncle Tom's Cabin* in 1851, publishing it bit by bit in *The National Era*, an antislavery newspaper based in Washington, DC. Despite the advance warning of serial publication, no one was prepared for the seismic impact of the finished book as it descended on the American nation in 1852, with North and South teetering on the edge of calamity, largely because of the target of her tale: human bondage.

But First Some Backstory

It's not obvious that a novel would possess such earth-shattering capabilities, but the form's history suggests the mimetic power to move and excite. Depending on how you define it, the novel goes back a couple of millennia in the Greco-Latin world, a millennium in Japan, and nearly as long in China. As for the first modern novel, the distinction usually falls to another of Stowe's childhood favorites: *Don Quixote*. Even in this primal specimen, Miguel de Cervantes constructs his story atop the notion that fiction can elicit responses of wild, outsized proportions.

The aged Don Quixote finds himself entranced by chivalric romances, fictional stories of questing knights on noble adventures. Worse than entranced, he's driven crazy. He spends so much time reading that he neglects his estate. "With too little sleep, and too much reading," says the narrator, "his brains dried up, causing him to lose his mind."

Fictional heroes and events ferment in Quixote's imagination, bubbling out in a new self-conception and mission in the world. "When his mind was completely gone," explains the narrator, "he had the strangest thought any lunatic in the world ever had . . . to become a knight errant and travel the world with his armor and his horse to seek adventures and engage in everything he had read." Characterization becomes identification, suggestion becomes action, and off he rides to joust with windmills and generally make an ass of himself. Of course, Cervantes wasn't against fiction any more than Plato actually opposed writing; that's why he gives us a novel of a quarter million words to tell us of the dangers.

Paolo and Francesca stop reading. Painting by William Dyce.
SOURCE: SCOTTISH NATIONAL GALLERY, WIKIMEDIA COMMONS.

This dynamic was long known and, like Quixote's romances, preceded the form of the novel itself. Christian hagiographies—saint's lives—were intended to evoke both devotion and imitation. When hearing, for instance, that Martin of Tours gave his cloak to a beggar, listeners to the account were supposed to discover not only feelings of admiration for the saint but also charity of their own welling up within, prompting similar acts of kindness and sacrifice from themselves.

It's worth saying the feelings stirred by stories might not always prove so edifying. One of the most famous scenes in Dante Alighieri's *Inferno* involves the doomed lovers Paolo and Francesca, caught and killed in adultery. "One day we read, to pass the time away," Francesca tells Dante—specifically, a romance of Lancelot and his affair with Arthur's queen, Guinevere, spurred on by the knight Galehot. "Time and again our eyes were brought together by the book we read," she explains. As they come to the place where Lancelot and Guinevere succumb to

temptation, so do they. "Our Galehot was that book and he who wrote it," she says. Through this passage and the romances to which it refers, the word *galeotto* slipped into Italian usage to signify a go-between, a procurer. "The pimp was the book," says Stanford's Robert Pogue Harrison.

If stories of any sort might have this kind of power, the novel amplified the potency, unleashing "torrents of emotion," as one eighteenth-century critic said of Jean-Jacques Rousseau's popular novel *Julie, or the New Héloïse*, and those emotions could have significant societal ramifications—wild and outsized—down to the fundamental reordering of norms and expectations.

"Witchcraft in Every Page"

Navigating those torrents of emotion, UCLA historian Lynn Hunt has argued novels set the stage for the expansion of human rights in the eighteenth century. The period witnessed a notable uptick in the publication of novels. When Cervantes published *Don Quixote* in two installments at the beginning of the seventeenth century, the reading public was still small, and novels were exceptionally novel. English examples include Richard Head's *The English Rogue* (1665); John Bunyan's *The Pilgrim's Progress* (1678) and *The Life and Death of Mr. Badman* (1680); and Aphra Behn's *Love-Letters between a Nobleman and His Sister* (1684–1687) and *Oroonoko, or The Royal Slave* (1688). But it's not like there were thousands more.

For the entire seventeenth century, the number of novels published across Europe amounted to hundreds. Then the tally jumped and gained momentum decade by decade throughout the eighteenth century. "In France," writes Hunt, "8 new novels were published in 1701, 52 in 1750, and 112 in 1789. In Britain, the number of new novels increased sixfold between the first decade of the eighteenth century and the 1760s: about 30 new novels appeared every year in the 1770s, 40 per year in the 1780s, and 70 per year in the 1790s." In England alone, we're now talking about thousands of individual titles in circulation by the start of the nineteenth century. The uprush in Germany was slower but eventually surpassed its rivals; in just one year, 1803, German publishers released more than 275 novels in time for Easter.

While a single novel might gobble up a family's fortnightly food budget, lending libraries made books more accessible, as did newspapers and journals that serialized novels. One Paris observer claimed, no doubt with some exaggeration, "Everyone, but women in particular, is carrying a book around in their pocket. People read while riding in carriages or taking walks; they read at the theatre during the interval, in cafés, even when bathing. Women, children, journeymen and apprentices read in shops. On Sundays people read while seated at the front of their houses; lackeys read on their back seats, coachmen up on their boxes, and soldiers keeping guard."

Describing the situation in Germany, a woman in 1784 wrote, "Here people are stuffed with reading matter in the same way geese are stuffed with noodles." Such reading matter included the Bible, the classics, newspapers, and, of course, newly fashionable novels.

The increase was not only notable; like anything new, it was also seen as potentially dangerous. Worried onlookers fretted about "reading mania" and compared books to narcotics and other addictive substances and compulsive practices. "No lover of tobacco or coffee, no wine drinker or lover of games," said one concerned clergyman, "can be as addicted to their pipe, bottle, games or coffee-table as those many hungry readers are to their reading habit."

Part of the worry involved concern that readers became emotionally swept up in their stories; this was all the more so, as Hunt notes, with the first-person and epistolary narratives popular at the time, in which the story unfolds through the direct speech and thoughts of the primary character or characters, a vantage point the reader vicariously adopts as if living their experience.

Looking at Samuel Richardson's *Pamela* (1740) and *Clarissa* (1747–1748) and Rousseau's *Julie* (1761), Hunt homes in on the reception and emotional impact the stories had on readers. The first, *Pamela*, concerns the story of the eponymous servant girl who fends off the lecherous advances of her wealthy employer, Mr. B. Readers went nuts for it. "It takes possession, all night, of the fancy," wrote one early reader quoted by Hunt. "It has witchcraft in every page of it; but it is the witchcraft of passion and meaning."

PAMELA;

OR

VIRTUE REWARDED:

IN A SERIES OF

FAMILIAR LETTERS

FROM A

BEAUTIFUL YOUNG DAMSEL TO HER PARENTS

PUBLISHED IN ORDER TO CULTIVATE
PRINCIPLES OF VIRTUE AND RELIGION IN THE
MINDS OF THE YOUTH OF BOTH SEXES.

BY MR. SAMUEL RICHARDSON

PAMELA
Mr B__. Pamela, and Miss Goodwin.

LONDON:
PUBLISHED BY J. S. PRATT.

MDCCCXLV.

Samuel Richardson's *Pamela*. SOURCE: BRITISH LIBRARY, ALAMY.

The tragic *Clarissa*, in which the eponymous heroine escapes an arranged marriage only to be raped by her rescuer and jailed in debtors' prison, produced similar results. Richardson's "Strokes penetrate immediately to the Heart," as one critic said, "and we feel all the Distresses he paints; we not only weep for, but with *Clarissa*, and accompany her, step by step, through all her Distresses." Said the poet Thomas Edward, "I never felt so much distress in my life as I have done for that dear girl." Another reader confessed difficulty sleeping. "My Spirits are strangely seized," she said. "I burst into a Passion of crying."

Rousseau's *Julie* had the same effect. Take just two examples Hunt provides, one from a future publisher and another from a retired military officer. "I have felt pass through my heart the purity of Julie's emotions," said the first. And the second: "You have driven me crazy about her. Imagine then the tears that her death must have wrung from me." That uptick in reading, the surging number of novels, the expansion of outlets for readers to encounter such works—it all goes back to an insatiable desire for more such vicarious experiences.

"A Technology for Perspective-Taking"

Little Harriet Beecher Stowe in the attic of her Litchfield, Connecticut, home could know and experience only so much. As she grew, she could learn what worked and what didn't, what to embrace and what to avoid by a mix of experimentation, observation, and reflection. But she'd be fooling herself if she thought that was all there was or that the sample size of direct experience was remotely sufficient.

"Why," asks philosopher Martha Nussbaum in her book *Love's Knowledge*, "can't we investigate whatever we want to investigate by living and reflecting on our lives?" Her answer is straightforward: "We have never lived enough." Our lives are too short and too specific—born, as each of us are, in a certain place at a certain time to certain parents with certain traits and certain prospects within a certain society and certain subcultures. Our perspectives are hemmed and constrained by the very factors that define us. We live in a box.

But, as young Stowe discovered, books offer a way out. Turning the pages of *The Arabian Nights*, she could follow Sinbad on his seven voyages, foil murderous thieves with Ali Baba's servant girl Morgiana, or cheer on Aladdin as he becomes fabulously wealthy, rescues the sultan's daughter, and kills a pair of wicked magicians—all with the help of not one but two genies!

As she read more mature and morally complicated books—say, one she recommended for others, Richardson's *Sir Charles Grandison*—she would be faced with dilemmas and decisions that prompted her own intuitions and instincts. Imaginatively, emotionally, ethically, Stowe would have stepped into roles—seen through the very eyes—of characters faced with trials and tribulations of all kinds. Will Harriet Byron succumb to

Hargrave Pollexfen's advances? Will Grandison fight Pollexfen in a duel? Will he honor his promise of marriage if it means renouncing his faith?

What would Stowe do? How would she react? As Hunt points out, the stakes in such questions were heightened by the narrative style of such novels as *Sir Charles Grandison*. Rather than merely observing the characters, readers participate in their lives, becoming intimates and confidants and, in a sense, becoming them. The nature of the story would prompt Stowe to identify with the plight of the characters, to empathize with them, to inhabit their experiences—hence the highly charged emotive reports of their original readers, highlighted earlier.

Books—especially stories, especially novels—offer a way out of the box. "Reading is," says Harvard psychologist Steven Pinker, "a technology for perspective-taking. When someone else's thoughts are in your head, you are observing the world from that person's vantage point. Not only are you taking in sights and sounds that you could not experience first-hand, but you have stepped inside that person's mind and are temporarily sharing his or her attitudes and reactions."

That's why we need fiction, says Nussbaum. It both expands the borders of the self and allows us to plunge deeper into that self. "Our experience is, without fiction, too confined and too parochial," she says, adding, "Literature extends it, making us reflect and feel about what might otherwise be too distant for feeling. . . . Literature is an extension of life not only horizontally, bringing the reader into contact with events or locations or persons or problems he or she has not otherwise met, but also, so to speak, vertically, giving the reader experience that is deeper, sharper, and more precise than much of what takes place in life."

Novels provide temporary leave of our specificity, an escape from the enclosed space of ourselves, the prison of our peculiar psychology. Literature brings us into the lives of others. If only for a few hours, we can appreciate their motivations and values; we can see what drives them, inspires them, and repels them. We can take the place of someone radically different from ourselves and engage the world as that self. As thriller writer Andrew Klavan once said, "It's intellectual sex."

"We demand windows," says C. S. Lewis in his epilogue to *An Experiment in Criticism*. "We seek an enlargement of our being. We want to be more than ourselves. . . . We want to see with other eyes, to imagine with other imaginations, to feel with other hearts, as well as with our own." As a tool, literature provides access to those other perspectives. And, importantly, the same dynamic also provides the possibility of moral instruction, persuasion, and improvement. After all, says Lewis, "every act of justice or charity involves putting ourselves in the other person's place and thus transcending our own competitive particularity."

Feeling with Very Other Others

Modern researchers have confirmed and explained this dynamic in several ways. For starters, reading a novel involves setting aside our own hopes and goals and assuming those of the principal characters. While we engage the story, we loan our cognitive and emotional processes to the characters. As our hero faces challenges, we problem-solve with them, guessing at the best possible responses given the option set; when their plans fail, we feel frustrated; when their enemies succeed, we feel dejected, anxious, angry, even vengeful; and when the hero finally prevails, we feel elated and satisfied, as if we have participated in their accomplishment. And, vicariously, we have.

These feelings are not fabricated or imaginary. By loaning our own cognitive and emotional resources to the character's purposes, we surface memories and generate feelings authentically ours. Some of these may have lain dormant and finally found cathartic expression. Others force reckonings with wrongs and injustices we'd likely never experience in real life but that activate our moral sense and cause us to reflect and judge in ways we rarely would in our day-to-day lives. "The emotions we experience," says University of Toronto cognitive psychologist Keith Oatley in a review of the scientific studies, "are not primarily those of the characters[;] they are our own, in the contexts we imagine." Fiction doesn't primarily happen on the page; it happens in our brains.

Such confrontations can have a transformative impact on our understanding of the world and even of ourselves. "Art," as Oatley says, "enables us to experience some emotions in contexts that we would not ordinarily encounter, and to think of ourselves in ways that usually we do not."

Stepping into the place of the characters facilitates reappraisal of our own beliefs and values, causing us to apply them in contexts we might not have previously imagined—and possibly revise them if they come up short, enlarging our moral horizons in either case.

There are valuable gains from this exercise. The real world, especially moments that involve our interactions with others, offers us a series of ethical tests, assessing the strength and subtlety of our emotional and moral intelligence. All else being equal, research shows that consumers of fiction perform better at these tests than do those who prefer other forms of entertainment. The "moral laboratory" of fiction serves as practice for the real thing.

Trained by the novels they read, consumers of fiction tend to exhibit sharper social awareness and greater ability to empathize. The remarkable fact of the eighteenth century? Thanks to the uptick in both readers and readers consuming novels, Hunt argues, these dynamics went to work on society at a scale never before seen and with characters rarely before considered, a development that would continue in the eighteenth century and beyond. The "technology for perspective-taking" suddenly included perspectives rarely taken before—people from all walks, including the poor and the marginalized.

By stoking empathy for the marginalized, novels expanded the moral awareness of their readers. Pamela, Clarissa, and Julie were all women in relatively powerless positions (indeed, Pamela was just a servant girl). And yet people of power and status could—and did—empathize with their plights, weep for their misfortunes, and desire redress for their wrongs. Perhaps this phenomenon is commonplace today; we can thank fiction, at least in part, and isolate its origins.

"The magical spell cast by the novel thus turned out to be far-reaching in its effects," says Hunt. "Human rights grew out of the seedbed sowed by these feelings. Human rights could only flourish when people learned to think of others as their equals, as like them in some fundamental fashion. They learned this equality, at least in part, by experiencing identification with ordinary characters who seemed dramatically present and familiar, even if ultimately fictional."

The highborn now peered through the lens of the low, forced to reevaluate their beliefs and values in the context of empathizing with those living in less privileged circumstances. And few found themselves in less privileged circumstances than did America's enslaved population.

The Desperate Birth of Uncle Tom

In 1850, the US Congress passed the Fugitive Slave Act, which renewed efforts to force runaways—men like James W. C. Pennington and Frederick Douglass—to return to their supposed masters. Signed by President Millard Fillmore as an attempt to placate angry enslavers in the South, the law stoked ire among abolitionists in the North, people who had been living alongside the self-emancipated and counted them as friends and neighbors.

As slave hunters wandered through cities like Boston and New York, they collared whomever they could. They encountered resistance. Some former slaves preserved their free status by running to Canada; others boarded ships to Europe. Some hid out in basements and attics and waited for the coast to clear—however long that might take. Abolitionists blocked enforcement of law when and however they could, facilitating escapes and interfering with slave hunters and the officials legally obligated to help them re-enslave productive citizens who had settled in their towns.

Abolitionists buzzed with righteous fury and frantic distress about the law and its effects. Harriet's sister-in-law Sarah, married to Harriet's brother George, wrote Harriet. Many in the family and their circle dedicated themselves to the abolitionist cause; Sarah and George's friend, the newspaper editor and Presbyterian minister Elijah Parish Lovejoy, was martyred by a pro-slavery mob in Illinois for publishing antislavery articles. Sarah suspected Harriet had something to offer the cause and penned several letters imploring her to act.

"I remember distinctly saying in one of them," she recalled many years later, "Now, Hattie, if I could use a pen as you can, I would write something that would make this whole nation feel what an accursed thing slavery is.'" Family members recalled Stowe sitting and reading the letter aloud with everyone listening. Finishing that desperate sentence, she stood, wadded the paper in her hand, and swore, "I will write something. I will if I live."

Stowe had written articles and some short stories, including one that garnered a prize from an organization called the Semicolon Club, but she'd never written anything so ambitious as *Uncle Tom's Cabin*. Still, she knew she must: Antislavery tracts had been written for decades and hadn't made the necessary difference. Like all those temperance and theological books and pamphlets in her father's attic library, nonfiction diatribes could do only so much. If she wanted to "make the whole nation feel," she needed to spin her own *Arabian Nights*, her own *Don Quixote*. Nobody was ready for it when she did.

"Trembling Every Nerve"

Stowe began portioning out passages of her novel, like Scheherazade sequentially spinning her tales in Stowe's beloved *Arabian Nights*. Starting in the summer of 1851, the serialization ran through the fall and winter, wrapping up in the spring of 1852. By then Stowe had sold the rights to a Boston publisher, who printed five thousand copies. Readers soon found themselves galvanized by the story.

The narrative opens on a Kentucky plantation, where Arthur Shelby, short on cash, plans to sell Uncle Tom and a little boy named Harry. The boy's mother, Eliza, hears of the plan and escapes north with him, risking the frozen Ohio River rather than lose her son; the woman's husband, George, had previously escaped, and she hopes to reunite their broken family. Meanwhile, Uncle Tom befriends little Eva St. Clare, the daughter of his new owner, aboard a steamboat headed south. Though his freedom is promised by Eva's father after he experiences a religious conversion, tragedy strikes before his emancipation, and Tom lands under the thumb of a brutal new owner, Simon Legree, who ultimately takes his life. Fortunately, events turn out better for Eliza and George, who not only find each other but also make their way north to Canada with the help of some Quakers.

By following the trail of her characters north and south, Stowe not only revealed the inhumanity of slavery as practiced below the Mason–Dixon Line but also indicted the Fugitive Slave Act, which mandated the merciless hounding of those who escaped the barbarity until they'd fled their own country, looking for refuge on foreign soil.

Informed by slave narratives, newspaper stories, and her own first-hand observation of life under slavery, Stowe's narrative gripped the nation. Her Boston publisher kept the presses running as demand soared through summer and fall. Read, reviewed, debated, praised, and scorned, the book went through printing after printing. Within the first twelve months, *Uncle Tom's Cabin* sold more than three hundred thousand copies, a number unheard of at the time. The book prompted a general upswell in antislavery publishing, including more novels.

As Stowe had hoped, readers confessed to being viscerally moved by the tale and the vicarious identification and suffering it invited. "We confess to the frequent misting of our eyes, and the making of our heart grow liquid as water, and trembling every nerve within us, in the perusal of incidents and scenes so vividly depicted in her pages," said the abolitionist William Lloyd Garrison, echoing the teary readers of *Pamela*, *Clarissa*, and *Julie* a century before.

"The touching, but too truthful tale of *Uncle Tom's Cabin*," said Frederick Douglass, who was then subject to the perils of the Fugitive Slave Act targeted by Stowe, "has rekindled the slumbering embers of anti-slavery zeal into active flame. Its recitals have baptized with holy fire myriads who before cared nothing for the bleeding slave." Stowe's novel, said Booker T. Washington, "so stirred the hearts of the northern people that a large part of them were ready either to vote or, in the last extremity, to fight for the suppression of slavery." Using this "technology for perspective-taking" and being able now to see with the eyes of the enslaved, sympathetic White readers were appalled by the vision—at least those in the North.

While Stowe hoped to appeal to Southern readers, many reacted with vitriol. Not only did Southern writers pen anti-*Tom* screeds, including some twenty-nine novels, but authorities also used the antiliteracy laws mentioned in the previous chapter to repress circulation of the book itself. One free Black minister in Maryland was caught with a copy and sentenced to ten years in prison for the supposed offense. Thankfully, his unjust sentence was later and decisively commuted by the Union army in a war Stowe's novel was said to have provoked.

"Is this the little woman who made this great war?" President Abraham Lincoln apocryphally said when meeting Stowe. But novels need not create effects so wild and outsized as wars to earn their place as essential tools of perspective-taking with the power to set nerves trembling and broaden the moral horizons of their readers. They do it to us and for us every day if we let them.

Marginalia: The Futility of Banning Books

The year 1982 witnessed not only the inaugural Banned Books Week but also the Supreme Court ruling in *Island Trees School District v. Pico*, a case in which the high court sided with students who sued their school district for removing books from their library. Then, as if to provide the ideal case study, 1982 further saw the publication of Alice Walker's *The Color Purple*—a novel that garnered for Walker the Pulitzer Prize, the National Book Award, and status as one of America's most banned authors.

Over the last thirty-odd years, the American Library Association's Office for Intellectual Freedom has compiled a list of the most banned books by decade. *The Color Purple* charts on every list: 1990–1999, 2000–2009, and 2010–2019. Attempts at banning began in 1984 when an Oakland, California, parent successfully petitioned her tenth grader's school to remove the book as "garbage." While the school board eventually relented and reinstated the book, *The Color Purple* has been targeted all over the country ever since.

Complaints include everything from Walker's dialect-infused narration to explicit language, drug use, sex—including sexual abuse and a same-sex relationship—physical abuse, and antireligious sentiment. In the mid-1980s, the primary gripe actually came from African American men complaining that Walker's depiction of domestic violence painted Black males as dangerous aggressors. The novel had something to offend almost everyone.

Of course, the book also won those coveted prizes and has sold more than five million copies, so it has its fans—often citing the same features that soured others. Oprah Winfrey, who acted in the movie, deeply identified with the book. "I opened the page and . . . I was like, 'There's another human being with my story,'" she told critic Salamishah Tillet. As an adolescent, Oprah had been raped several times by male relatives and family friends, all of it, including the death of her premature baby, hushed up and ignored.

Oprah discovered the book through a *New York Times* review while living in Baltimore, Maryland. Still in her pajamas, she left the house to purchase a copy from a nearby bookstore and read it in a single sitting. She then bought up every copy she could find and began handing them out to everyone she knew. It was, said Oprah, "the single most defining experience I've ever had." Had she not stumbled onto that review, she said, her "life would have gone a completely different direction."

What in this story moves people both for and against? Walker set her tale in rural Georgia in the decades between World Wars I and II. The narrative moves through a series of letters, first from the heroine Celie to God, unburdening her mind and heart about her life, beginning with her rape at fourteen years old by her stepfather. Told to keep quiet, the young girl never betrays him, though she finds herself with child twice; her stepfather takes the children away—she assumes to kill them.

Eventually the man tires of Celie and marries her off to another. This man goes by the name Albert, but Celie simply refers to him as "Mr. _____" in her letters. Similarly abusive, Albert beats Celie to keep her in line. What's more, he invites his mistress, blues singer Shug Avery, to stay in their home. And, worst of all, he hides letters sent from Celie's little sister Nettie, who attaches herself to a missionary couple en route to Africa and knows the true fate of Celie's two children.

Perhaps it's little surprise that Celie, exploited by men, feels no affection for them. The surprise? She forms a friendship with her husband's mistress that eventually turns romantic. Shug still carries on with Albert and, after she leaves and marries a man named Grady, continues her relationship with Celie. In time she convinces Celie to leave Albert over his betrayal with the letters and to come with her to Memphis to start a tailoring business.

Through these ordeals, Celie stops writing to God, convinced he's done nothing for her. But her tragic life does turn for the good. With Shug and Celie gone, Albert softens and even reforms; when years later Celie returns to inherit property stolen by her now-deceased stepfather, she finds herself able to befriend Albert. And it's there that she's finally reunited with her sister Nettie and her two lost children.

It's a brutal tale that perhaps pushes more panic buttons for some readers than others. But need it? Take the claim of the book's antireligious message. Walker presents Celie's early letters in the book as a sort of failed theodicy, but Nettie remains a faithful believer, marries a missionary, and emerges as one of the novel's many uplifting elements. What Celie explicitly denies is the idea that God is a disappointing, barefooted, gray-bearded White man in a robe. It's hard to imagine anyone storming the battlements for such a caricature.

But storming battlements is predictable behavior during the sort of culture-war skirmishing that book bans and challenges represent. While various sides contest the status of a few thousand books every year in school districts across the county, no one can agree on criteria for inclusion or exclusion. Meanwhile, the very act of fighting over individual titles tends to raise their profile and fuel their sales, something the promoters of Banned Books Week gleefully recognize.

Should teenagers be required to think about such issues as violence and sexual abuse? Novels dealing with difficult material represent one way of processing these sorts of subjects. It's why high school reading lists have long featured challenging books such as Walker's, Harper Lee's *To Kill a Mockingbird*, Joseph Heller's *Catch-22*, Zora Neale Hurston's *Their Eyes Were Watching God*, and Ken Kesey's *One Flew over the Cuckoo's Nest*.

"Book bans inhibit a core function of public education," says constitutional attorney David French, writing in *Reason* magazine. "They teach students that they should be protected from offensive ideas rather than how to engage and grapple with concepts they may not like." In making his case, French points back to the *Island Trees* case, in which students argued that denying them access to information undermined their First Amendment rights. The court agreed—in part because the ban would impair their development as citizens.

"Just as access to ideas makes it possible for citizens generally to exercise their rights of free speech and press in a meaningful manner," ruled the court, "such access prepares students for active and effective participation in the pluralistic, often contentious society in which they will soon be adult members." In other words, if students can't handle uncomfortable content in a book, what makes us think they can handle the rough-and-tumble world of democratic debate and civic engagement?

Chapter 15

BROWSING THE UNIVERSAL LIBRARY

Information Overload

A tool is a response to a problem, a remedy for natural human deficiency. In 1927, for instance, pioneering engineer Vannevar Bush designed an analog computer for solving high-dimension differential equations that mathematicians, physicists, and engineers had previously found brutally difficult to solve. "'Thinking Machine' Does Higher Mathematics," declared a *New York Times* headline when covering the story; "Solves Equations That Take Humans Months."

Throughout the 1930s, Bush became increasingly aware of another problem, one bedeviling researchers in all fields. "We are becoming bogged down," he said in 1939. "There is a growing mountain of research results; the investigator is bombarded with the findings and conclusions of thousands of parallel workers which he cannot find time to grasp as they appear, let alone remember."

As scientific work expanded, specialized, and accelerated, the growing literature outpaced anyone's ability to manage. A person could read twelve hours a day and never get to the end of it. There was simply too much to know, much less keep up with. And yet, said Bush, "we adhere rather closely, in our professional efforts, to methods of revealing, transmitting, and reviewing results which are generations old, and now inadequate for their purpose."

It might surprise us to think of *information overload* as a pressing problem before the digital era; the term itself only came into vogue after the 1960s. But the problem began presenting itself long before.

Both Cheap and Expensive

In July 1876, America marked its centennial with an exhibition in Philadelphia. Various technological marvels demonstrated the adolescent nation's march of progress. Participants could, for instance, see a press owned by Benjamin Franklin capable of printing 150 sheets of paper in an hour. In Franklin's day, as in Gutenberg's, printers fed one sheet at a time into a flat press cranked by hand. But a lot can happen in a century. In the intervening years, innovators powered presses with steam, pressed type on paper with rolling cylinders, and fed paper through the rotors, not sheet by sheet but in a continuous stream from giant rolls. Inventor Richard Hoe spearheaded many of these improvements; after observing Franklin's humble press, exhibition participants in Philadelphia could turn and see one of Hoe's machines in action, spitting out thirty-two thousand complete copies of a newspaper in just sixty minutes.

Industrialization radically altered the media landscape. Not only did it drop the price of print, but it also deluged readers with material. As we saw earlier, the flood of information created challenges during the first days of print. Now it overran almost any hope of containment. But just as enterprising scholars and librarians faced fresh difficulties in the early modern era with novel organizational solutions, in the wake of industrialization scholars and librarians looked for new ways of managing the overflow of information. Their work would eventually lead to the personal computer, the World Wide Web, and generative artificial intelligence (AI)—especially large language models.

At every step and stage, these innovators had to recalculate the cost of finding and deploying information. If, as economist Tyler Cowen said, "once an idea has been generated, it can be used many times by many different people at very low marginal cost"—an observation we've highlighted before—it's true only insofar as tools for accessing that idea can keep pace with the production of all other ideas. If, however, as Vannevar

Richard Hoe's web press at the centennial exhibition in Philadelphia, capable of printing thirty-two thousand newspapers in an hour. *SOURCE: HARPER'S WEEKLY*, DECEMBER 9, 1876. AUTHOR'S COLLECTION.

Bush warned, the literature relating to a certain topic grows beyond one's ability to locate a particular fact, then the marginal cost for using that idea is no longer low; the only way to tamp down the cost is to increase the discoverability of individual ideas.

The ancient invention and use of metadata exemplifies successful efforts to improve discoverability. From grouping and labeling clay tablets to tagging scrolls with titles, archivists, librarians, and scholars devised ways to quickly identify desired texts. Later tools such as indexes and concordances further enabled readers to burrow through books for key passages. But what happens when the total volume of information outstrips even the most advanced systems' ability to locate a choice bit of data, a juicy anecdote, or an obscure quote or insight?

"An Inventory of All That Has Been Written"

As a boy, the future librarian Paul Otlet devised his own catalog for his personal books and papers. Born in Belgium in 1868, Otlet was appointed librarian of his Jesuit school at just sixteen years old. His responsibilities included helping students find the books they needed and keeping the shelves in shape following the order in the library catalog.

"It seemed a wonder," Otlet said of the catalog, "this instrument that allowed me to use all of these books." Of course, it was exactly this kind of catalog that would soon prove ineffective against the onrush of industrialized publishing. There's a half-life to every solution, but, conveniently enough, Otlet was also the kind of person who could imagine what the new circumstances required.

The first step? Eliminate the extraneous. "The ideal," Otlet explained, "would be to strip each article or each chapter in a book of whatever is a matter of fine language or repetition or padding and to collect separately on cards whatever is new and adds to knowledge." Excluding fiction and other works of art, knowledge could be reduced to one or more of four key components: facts, interpretations, statistics, and sources. Otlet first conducted this procedure by cutting bits and pieces out of books and articles and pasting them to index cards; later he adopted a typewriter for the job. He essentially unbound the book and extracted its contents for a different mode of engagement.

The next step was to overhaul library classification. A traditional catalog could take a reader as far as an individual title, where the book's table of contents and index had to provide additional help in searching. Otlet wanted something more robust. Now that he'd atomized books and articles for individual facts, Otlet created a catalog that could identify the bits and pieces—paragraphs, sentences, and keywords. As impressive as that is, however, creating an analog Google was only half the advance.

Traditional library classification follows a hierarchical structure. If you, for instance, wanted a book about English slang, the Dewey Decimal System would lead you down a linear path, starting at languages (400s), narrowing to English (420s), and finally landing on English language variations (427). But you could imagine arriving by other paths—say, various historical, regional, or ethnographic studies. Knowledge doesn't fit in a bento box; it exists in multiple categories simultaneously. Breaking free from simple hierarchies, Otlet realized he could tag a fact with not only time periods and geographical locations but even additional topics. Categorization could suddenly work laterally as well as linearly.

Working throughout the 1890s with his business partner, Henri La Fontaine, Otlet began building what amounted to a massive analog data-

Paul Otlet, immersed in information—like all of us. *SOURCE:* ART COLLECTION 4, ALAMY.

base and search engine. When the pair debuted their creation at the Paris World Fair in 1900, they'd already produced more than three million index cards, a fifth of their eventual total, a gargantuan book bound not in covers but in miniature filing cabinet drawers. Their goal was to assemble, as Otlet put it in 1897, "an inventory of all that has been written at all times, in all languages, and on all subjects."

They failed. Competing projects, insufficient funds, shifting state support, and two world wars frustrated the pair's efforts. Still, Otlet realized the challenge posed by an abundance of information and left a model for handling the mounting burden. As his vision developed over the decades, he imagined a vast multimedia network, combining his massive databank with telephones, television, radio, phonographs, and film—an electromechanical internet. Researchers could submit queries for a fee and receive answers to their questions, efficiently culled from one or more of Otlet's millions of index cards.

The First Home Computer

Though Paul Otlet and Vannevar Bush shared at least one acquaintance in common, Bush seems to have developed his solution to information overload without direct reference to Otlet's plan. While Bush noodled and tinkered on it throughout the 1930s, he finally went public with his solution in "As We May Think," a landmark article in the July 1945 *Atlantic Monthly*, later abridged and published by *Life*, reaching a combined readership in excess of two million, not to mention wider coverage by the Associated Press and *New York Times*.

Echoing his 1939 statements, Bush mentioned "a growing mountain of research," which leaves knowledge workers "staggered." For knowledge to be valuable, it has to be consulted, extended, and stored for the next round of consultation. Too much information slows or stops the process.

The cost? "Mendel's concept of the laws of genetics," said Bush, providing one example, "was lost to the world for a generation because his publication did not reach the few who were capable of grasping and extending it; and this sort of catastrophe is undoubtedly being repeated all about us, as truly significant attainments become lost in the mass of the inconsequential." Further, as Bush noted, this problem goes well beyond scientific research: "It involves the entire process by which man profits by his inheritance of acquired knowledge."

In an earlier essay, "The Inscrutable Thirties," published in 1933, Bush provided a picture of the logistic hassle involved in working amid so much information. He asked readers to picture a professor in a library: "Long banks of shelves contained tons of books, and yet it was supposed to be a working library and not a museum. He had to paw over cards, thumb pages, and delve by the hour. It was time-wasting and exasperating indeed." Bush thought he could lessen the difficulty by condensing all those books on microfilm so that "the content of a thousand volumes" could be consulted "by depressing a few keys" and having "a given page instantly projected before him."

Where would this microfilm be stored? How would the desired passages be called up? On what would it be projected? As Bush explained in "As We May Think," he envisioned a desk that contained reels of microfilm, a keyboard, and an array of buttons and levers. These would allow the user to input codes to trigger a mechanism to rapidly search the relevant data on the microfilm, be it books, reports, business communication, personal records, whatever. Two slanted screens on the desktop would then serve up the selected content, and the levers would facilitate manipulating the documents on the screens. He called this theoretical device the Memex, essentially an analog personal computer. (Otlet imagined a similar workstation called the Mondotheque.)

Bush saw the Memex as a way not only to rapidly sort and consume content but to add to it as well. A user could augment their library by either scanning documents or inserting preloaded microfilm. They could likewise add their own long-hand annotations with a stylus and even connect documents together in user-specific associations (prefiguring both hypertext and hyperlinks). Bush called these connections "associative trails" and saw them as vast improvements over conventional alphabetical or numerical indexing.

Unlike linear connections between concepts, an associative trail could link items together more flexibly, moving between subject areas, even entire fields, joining all manner of inputs and diverse media—text,

Vannevar Bush (middle) yukking it up with nuclear physicists. *SOURCE:* EVERETT COLLECTION HISTORICAL, ALAMY.

images, audio recordings—into novel configurations. "It is exactly as though the physical items had been gathered together from widely separated sources and bound together to form a new book," said Bush. Otlet had a similar idea, referring to these connections with a term we would use—each one a *link*—and seeing the sum of these links forming a new, emergent "Universal Book."

"Bush's great insight was," said Brewster Kahle, founder of the Internet Archive, "realizing that there's more value in the connections between data than in the data itself." But that's really only part of the story. Though Bush's Memex and its theoretical underpinnings rested on analog technologies, his thinking rolls off in two very contemporary, complementary digital directions—namely, the internet and AI large language models.

Network Effects

A key limitation of storing information in physical formats is their dependence on location. Molecules existing in one place cannot exist in other places at the same time. A particular book or file in Boston is inaccessible to someone in San Francisco, Chicago, or New York—or anywhere else, for that matter—without being physically transported from one place to another. Books are "portable magic," as Stephen King said, but a more potent magic would be omnipresence.

Vannevar Bush recognized the problem. His explanation of the Memex in 1945 described a self-contained local device. New files could be added but only manually. Given the value of "associative trails" through research, however, that presented a major drawback.

Imagine subject matter experts in, say, chemistry, law, history, or biology creating helpful trails through the specialized literatures of their respective fields; wouldn't other users want to access those expertly blazed trails? Bush continued to develop his idea, and his notes reveal a fruitful path forward. "Professional societies will no longer print papers," he said in an unpublished 1959 manuscript called "Memex II." "If [a researcher] is in a hurry, he can dial a telephone call to the society and then dial further to identify a paper, whereupon it will be transmitted over his phone connection and entered into his record immediately and directly, by facsimile transmission."

This idea is similar to querying Otlet's library of index cards, but the data transfer would be automated through the phone lines by entering call numbers. Bush is describing a peer-to-peer file-sharing network between Memex devices—again, all analog. Others were thinking along similar but more grandiose lines, especially those already working with digital machines.

While serving as the director of the Information Processing Techniques Office (IPTO) at the Advanced Research Projects Agency (ARPA), J. C. R. Licklider imagined linking geographically separate mainframe computers together in a network. His ambitions ran high. In 1963, he placed tongue in cheek and sent an interoffice memo addressed to "Members and Affiliates of the Intergalactic Computer Network." The memo referenced the basic difficulty solved by a network: Information and programs stored in one locale could instantly be "brought into the part of the system that I was using." The limits of location were thus all but eliminated; information went from bounded and scarce to ubiquitous.

Ahead of his time, Licklider saw computers as something other than mere calculating machines; more fundamentally, he saw them as communication devices. He argued as much in an influential 1968 paper written with his protégé Robert Taylor. As Licklider's successor at IPTO, Taylor created ARPANET, the precursor to the internet. Through the subsequent work of Doug Engelbart, Ted Nelson, and Tim Berners-Lee (all of whom revered Vannevar Bush's 1945 Memex article), we later got full-blown hypertext and the World Wide Web.

In his initial proposal in 1989 for what became the World Wide Web, Berners-Lee specifically cited the challenge of discoverability as the reason for supporting a network of hyperlinked media. "Often, the information has been recorded," he said; "it just cannot be found." The solution? "A web of notes with links" that "could grow and evolve."

Of course, the eventual proliferation of hypertext documents on the web created discoverability problems of its own, something Berners-Lee anticipated. "An intriguing possibility, given a large hypertext database with typed links, is that it allows some degree of automatic analysis," he said. "This is particularly useful when the database becomes very large, and groups of projects, for example, so interwoven as to make it difficult to see the wood for the trees." Automatic analysis? What Berners-Lee is talking about is artificial intelligence. Interestingly, both Vannevar Bush and J. C. R. Licklider were already there.

Machines for Thinking

In "As We May Think," Bush conjured a scene in a laboratory. A researcher moves around his work, observing an experiment. With free hands, he photographs what's happening and audibly comments on the unfolding display. A machine captures his images, records his voice, and time stamps both to sync the output. A radio connection allows him to observe and record when he's working in the field. At the end of the day, he then reviews his notes, thinking and rethinking about what he's observed, adding to his reports—still doing so by simply speaking aloud, his Memex autocompiling the audio recording, photos, and notes.

Still, there's more work to be done. "Much needs to occur," said Bush, "between the collection of data and observations, the extraction of parallel material from the existing record, and the final insertion of new material into the general body of the common record." Some of that work would, of course, require original thought and synthesis. But some of it would be rote and even—we might intuitively employ a metaphor here—mechanical. So, why not a machine? "For mature thought there is no mechanical substitute," said Bush. "But creative thought and essentially repetitive thought are very different things. For the latter there are, and may be, powerful mechanical aids."

From time immemorial we humans have, as Bush observed, used tools to extend the potential of our native capacities. We developed machines to multiply our abilities, augment our strength, enhance our vision, reconfigure our resources, and traverse distances impossible without technological assistance. "Now man takes a new step," said Bush. "He builds machines to do some of his thinking for him."

Of course, as mentioned, Bush had already taken that step. He developed and employed his Differential Analyzer to solve exceedingly complex equations with multiple variables. While the analog computer was later retired in favor of more advanced digital systems, Bush's machine enabled, as an example, US forces in World War II to compute the precise trajectories required to successfully bomb Nazi targets, factoring such variables as gravity, winds, and projectile velocity.

The Analyzer performed these calculations far more efficiently and reliably than humans. But more to the point, "it demonstrated," says Belinda Barnet of Swinburne University of Technology, "that machines could automate human cognitive techniques." And Bush suspected this ability could go far beyond math: "Whenever logical processes of thought are employed—that is, whenever thought for a time runs along an accepted groove—there is an opportunity for the machine. . . . We may some day click off arguments on a machine with the same assurance that we now enter sales on a cash register."

As Bush's plans for the Memex evolved between the 1930s and 1960s, he saw the device shouldering increasingly more of its user's cognitive load. He imagined the Memex observing its user's preferences and updating its memory accordingly, particularly the user's unique "associative trails" through his research, saving him the mental bandwidth of processing them for himself. He also imagined the user could assign research to the Memex, which it could then perform while he stepped away—essentially creating dossiers and briefs in the user's absence.

Bush envisioned further uses, including more complicated cases requiring extensive analysis and judgment in, for instance, medicine, law, history, and the sciences, all of which would permit the human to perform at a higher level. Recalling Cicero's use of the Latin term *mens*, Bush's machine had a form of intelligence, and the human user could harness that ability to enhance his own. If books were tools for thinking, then the tools that helped people use them better were as well.

The Library and the Librarian

Citing the flood of print and its consequent challenges, in 1956 the Ford Foundation established the Council of Library Resources to modernize cataloging, retrieval, and the wider range of functions necessary for researchers to fully benefit from the libraries they used. The council sponsored a study led by J. C. R. Licklider in the two years between November 1961 and November 1963, resulting in Licklider's 1965 report, *Libraries of the Future*.

In this report, Licklider noted that the printed page "is superb" for displaying information but that the codex presented several drawbacks, all the more so when collected in a library. While for its original users the codex represented a marked improvement over the scroll, and the scroll over the clay tablet, the deluge of print complicated the use of the reigning format, as Otlet and Bush had already argued. But Licklider pushed the complaint further than either of his predecessors.

It's worth quoting him at length. "We may seek out inefficiencies in the organization of libraries," he said,

J. C. R. Licklider at the center of the intergalactic computer network. *SOURCE:* PICTORIAL PRESS LTD, ALAMY.

but the fundamental problem is not to be solved solely by improving library organization at the system level. Indeed, if human interaction with the body of knowledge is conceived of as a dynamic process involving repeated examinations and intercomparisons of very many small and scattered parts, then any concept of a library that begins with books on shelves is sure to encounter trouble. Surveying a million books on ten thousand shelves, one might suppose that the difficulty is basically logistic, that it derives from the gross physical arrangement. . . . [I]n much greater part the trouble stems from what we may call the "passiveness" of the printed page. When information is stored in books, there is no practical way to transfer the information from the store to the user without physically moving the book or the reader or both. Moreover, there is no way to determine prescribed functions of descriptively specified informational arguments within the books without asking the reader to carry out all the necessary operations himself.

In libraries like those of Plato, Aristotle, or Eusebius, that was doable. Even in swelling post-Gutenberg libraries, the difficulty could be met by new organizational tools, such as those pioneered by people like Conrad Gessner and Hernando Colón. But in a world of practically infinite publication, the old arrangements, methods, and techniques came up short. For Licklider, that would never do; the council that hired him likely knew where his solution would tend.

Five years before his final report for the council, Licklider published a landmark essay of computer science, "Man-Computer Symbiosis," in which he argued those "necessary operations" ought rarely to be left to the human alone. The goal, he said, was "to think in interaction with a computer in the same way that you think with a colleague whose competence supplements your own." What's more, the computer would help the human user frame his queries and problems and work in real time on the solutions. No lag—more like another person in the room than a research assistant who meanders off to work on their own and present their results at a later interaction.

To accomplish this goal, people would need to think beyond the codex—indeed, beyond the concept of texts. Instead of documents, Licklider proposed atomizing books down to discrete blocks of information, exactly like Otlet had suggested more than a half century before but now assisted by digital tools, not just paper index cards and typewriters. (I should add: Also like Otlet, Licklider was referring to data, not art; Shakespeare, Austen, and Hemingway would remain untouched in this scheme.)

The important thing is that once computers had digested all this data, they could serve it back to researchers, reconfigured, updated, and elaborated as needed. We're well beyond the traditional library at this point. In fact, Licklider suggested a coinage with zero chance of adoption but that captured his ambitions perfectly; he recommended calling computerized libraries "procognitive systems" in that the library would actively contribute to the cognitive deliberations of its users. No more "'passiveness' of the printed page."

The proaction of the system was a key feature for Licklider and echoed Bush's eagerness for the Memex to automatically serve up relevant "associative trails" for its user. Amid an information glut, in which a codex's physical features are a bug, not a feature, and catalogs and indexes are insufficient to the task of searching and finding relevant data, knowledge workers waste an inordinate amount of time, as Licklider said, "searching, calculating, plotting, transforming, determining the logical or dynamic consequences of a set of assumptions or hypotheses, preparing the way for a decision or an insight." After studying his own practice, he determined "about 85 percent of my 'thinking' time was spent getting into a position to think."

If a machine could shortcut the more clerical or functionary tasks, such a system would utterly transform our interaction with recorded knowledge. "A basic part of the overall aim for procognitive systems is to get the user of the fund of knowledge into something more nearly like an executive's or commander's position," explained Licklider in a passage that remarkably prefigures the way people use large language models today. "He will still read and think and, hopefully, have insights and make discoveries, but he will not have to do all the searching himself nor all the transforming, nor all the testing for matching or compatibility that is involved in creative use of knowledge. He will say what operations he wants performed upon what parts of the body of knowledge, he will see whether the result makes sense, and then he will decide what to have done next."

In other words, the human conceives of a prompt and then evaluates the results; the machine handles the grunt work, leaving the more theoretical work to its operator. Sitting atop the internet—the Universal Library that Paul Otlet envisioned—AI is both library and librarian. And this may be the least appreciated feature of large language models.

Evolution of the Idea Machine

Ever since humans first domesticated data with the advent of writing, our accumulation of information has posed challenges—storing, searching, sorting, editing, utilizing. Large language models offer a radical solution to these difficulties. Of course, as with any solution, there are trade-offs, and it may take us time to fully appreciate those, just as it will take us time to develop workarounds and compensations.

As it turned out, Paul Otlet's dream came to nothing, and Vannevar Bush never built the Memex; the necessary technology was unavailable at the time. But the visions of both men live on in the World Wide Web and generative AI. It was, in a sense, inevitable. AI is embedded in the logic of the internet, and the internet is embedded in the logic of libraries, and libraries are embedded in the logic of the book, and the book is embedded in the logic of tools, and tools are embedded in the logic of the human imagination.

Marginalia: Back to the Beginning

The earliest chapters of this book explored some of the earliest chapters of human history, but much of that history remains obscure, veiled by the very texts that might shed light on the past and yet, for a variety of reasons, remain undeciphered.

Consider cuneiform, the first script in which humans recorded our business, our myths, our poetry, our hymns, our history. Hundreds of thousands of recovered cuneiform clay tablets representing more than a dozen different languages and detailing aspects of ancient life and culture remain untranslated to this day. The trouble? Fewer than a hundred people living in the world today can fluently read the script. What if we could get those humans some help?

For many years, a multidisciplinary team of Ancient Near East experts, digital humanists, and computer scientists based at the University of Chicago has been working to decipher forty-five-hundred-year-old cuneiform documents from the Achaemenid Empire, written in the Elamite language. To successfully translate these tablets, scholars must pick through a hundred thousand individual wedge-shaped signs, identify their meaning, transliterate them phonetically into Latin letters, and then translate those transliterations into English. It's a painstakingly manual process—unless a generative AI model can shoulder some of the work.

Using five thousand images of cuneiform tablets, the team trained their model to identify individual characters and provide scholars with transliteration suggestions. And it did so far faster than humans, with greater accuracy and consistency.

Another team of researchers, this one from Israel, recently built an AI model to help carry Akkadian cuneiform texts over the linguistic chasm into English. The machine could handle both transliteration and direct translation from Akkadian, performing best with short and medium-length sentences and with straightforward genres, such as letters and official records. While the model sometimes hallucinated and struggled with more figurative language and complicated texts, it still served up sufficiently accurate translations almost instantaneously, freeing human scholars to spend their scarce time improving the translations and interpreting and applying the newly unlocked information.

"Translating all the tablets that remain untranslated could," said project leader Gai Gutherz, "expose us to the first days of history, to the civilization of those people, what they believed in, what they were talking about, what they were documenting." And other periods of history can be similarly illumined. Millions of medieval manuscripts exist in museums, monasteries, and other research libraries; some 160,000 reside at the Abbey Library of St. Gall in Switzerland alone, texts going back to the eighth century—before Charlemagne's reforms.

Using digitized images of such manuscripts, an interdisciplinary team at the University of Notre Dame has developed an AI model to facilitate transcription and translation of these documents. "There is all sorts of information hidden in these manuscripts," said Hildegard Müller of Notre Dame—"unidentified texts that nobody has seen before."

Though they began with Latin texts, scholars have since updated the model to read Ethiopian manuscripts as well. And transcription and translation represent just the beginning of the benefits because an AI model that can transcribe scripts from the eighth, ninth, tenth, and eleventh centuries can also render them searchable to an extent never seen in any century—a mix of Robert Grosseteste's index and Hugh of St. Victor's concordance with the dynamism and scale of which neither could ever dream.

Because of AI, secrets of the past have never had a greater chance of being uncovered. Some cases, such as efforts with Elamite, Akkadian, Latin, and Ge'ez, represent astounding labor-saving achievements; in other instances, AI is able to do what even entire armies of humans couldn't.

What if, for instance, you had access to a Roman library like Cicero's? Not a reconstruction of a collection that he or one of his contemporaries might have owned, stocked with scrolls copied and recopied over two millennia and inevitably corrupted by time and human error. No, what if you had the original books themselves?

When Vesuvius blew its top in 79 CE, the town of Herculaneum was scorched by volcanic gas and buried under debris. The town was unearthed sixteen centuries later, beginning in 1709; the find included a villa scholars assume belonged to Julius Caesar's father-in-law. And inside? More than a thousand scrolls, the largest intact library from the classical past—maybe *too* intact, actually. Much of the collection was instantly carbonized by the heat of the eruption, and the scrolls were turned black as coal and fused shut.

Early efforts to unroll the books caused them to flake into fragments. Just as tricky, the scrolls can't be read with the naked human eye. Computer tomography and multispectral photography do allow researchers to peer through the rolled layers of papyrus and even identify ink, but in most cases they require more advanced technology to decipher the content of the scrolls.

Previous solutions made some progress, but, not surprisingly, AI has provided the best path forward here as well. By early 2024, more than two thousand characters forming fifteen individual passages had been read from an unrolled scroll through a combination of computer tomography and AI. Once the model virtually unrolled the scroll and identified the script, a team of papyrologists could assess the writing and read words frozen in time for nearly two millennia—a digital window onto the analog past.

REFLECTIONS

The technologies we create also shape the ways that we think and act in the world, and this, in turn, influences the kinds of technology we further invent and use.

—Robert Hassan

And now the machine is part of us, like our arms and legs—more important than either, for we couldn't even live if the machine were amputated from civilization.

—John Wyndham

Chapter 16

ENGINES OF CHANGE

"If You Wrote a Book, You F—d Up"

Before the downfall of Sam Bankman-Fried, founder of FTX, once the world's third-largest cryptocurrency exchange, people regularly pointed to him as a sage and moral exemplar, owing to his embrace and advocacy of effective altruism. But his extreme utilitarianism pushed him in some unfortunate directions.

"I would never read a book," he told journalist Adam Fisher in an interview for a lengthy profile originally published on the website of FTX investor Sequoia Capital in 2022.

"For me," said Fisher in the exchange, "reading books is the highest-bandwidth way I know to get quality information into my brain, which just craves the stimulation." Not for Bankman-Fried.

"I'm very skeptical of books. I don't want to say no book is ever worth reading, but I actually do believe something pretty close to that," he said. "I think, if you wrote a book, you f—d up, and it should have been a six-paragraph blog post."

Less than two months after the profile ran, FTX unraveled. Bankman-Fried resigned, and the exchange filed for bankruptcy. Less than two years after that, a US district court found Bankman-Fried guilty of defrauding customers and investors of $10 billion. He was sentenced to twenty-five years in prison.

Bankman-Fried's dismissal of books was so emphatic and published so close to his downfall that it became a theme in the analysis of his demise. "It's impossible to read the sad saga of Mr. Bankman-Fried without thinking he, and many of those around him, would have been better off if they had spent less time at math camp and more time in English class," said *New York Times* reporter David Streitfeld. "Sometimes in books, the characters find their moral compass; in the best books, the reader does, too."

Streitfeld's observation harkens back to one of the benefits of books we explored earlier: the moral enlargement facilitated by such books as *Clarissa* and *Uncle Tom's Cabin*. And the uses extend far beyond calibrating one's moral compass. As we've seen, Microsoft cofounder Bill Gates is a reader; the same is true for current Microsoft CEO Satya Nadella, Amazon founder Jeff Bezos, and many, many others of similar accomplishment, including techno-optimist and *Whole Earth Catalog* founder Stewart Brand. They all find books a manifestly useful technology for a variety of ends.

Brand's enthusiasm for the book is central to his conception of the Long Now, which imagines human civilization extending another ten thousand years. Part of the project includes the Manual for Civilization, a crowd-curated library of thirty-five hundred books necessary to restart civilization from scratch should circumstances so dire ever arise.

Bankman-Fried could have learned something from such examples. Instead, his story stands as an accidental defense of the humanities— making him a reverse poster boy for the real value of an English degree. "I would never read a book"? Maybe he should have. Anyone who dismisses a technology without understanding its potential or its uses is a fool.

The Bias of Books

Technologies, as media theorist Douglas Rushkoff says, have biases. They suggest some uses more than others. "People like to think of technologies and media as neutral and that only their use or content determines their effect," he says. "Guns don't kill people, after all; people kill people. But guns are much more biased toward killing people than, say, pillows." If we fail to recognize the biases of the tools we use, our tools will, as Rushkoff says, "confound us," much as books evidently confounded Sam Bankman-Fried.

The idea machine displays a bias like any technology. I'd like to suggest several traits that stand out to me from the foregoing history.

First, because they're built on a foundation of writing, books enable ideas to be developed, defined, and elaborated beyond the capabilities of the unassisted human mind. As we've discussed, writing moves ideas along the x- and y-axes of the idea grid, amplifying both their expression and their construction; meanwhile, books enable ideas to travel along the z-axis, enabling well-formed ideas to persist through time. Ideas conceived in one generation become not only accessible but also open to new interpretations and applications in another, centuries and even millennia later. We still find uses for Moses, Plato, Aristotle, Paul, Augustine, Aquinas, Machiavelli, and countless other voices from the past.

Second, because books allow ideas to persist through time, ideas accumulate and expand, with books serving as ever-rising cultural reservoirs or interest-compounding societal savings accounts. That is to say, the idea machine powers cultural growth and societal evolution, with new books contributing to a ratchet effect, inching, nudging, budging, and facilitating the next possible path.

Sam Bankman-Fried should have read a book.
SOURCE: JOHN ANGELILLO, UPI/ALAMY LIVE NEWS.

I hesitate to say "path forward" since that presumes an end goal that does not exist. Progress is a story we tell retrospectively, and if we look back, we can see books directly serving the advances we've made, not only as individual societies but also as a species. Gifted with a mind of certain capacities, we leveraged the idea machine to expand its capabilities, and the idea machine has enabled and even driven the shape and trajectory of what we regard as human progress.

Third, the idea machine is practically self-organizing. That is, it suggests ways of accessing and using the ideas we employ it to generate. This quality is evident in the libraries we've explored, those of Montaigne, Sidonius, Eusebius, Cicero, Aristotle, Plato, Ashurbanipal, and others. Metadata emerges as we use data. Looking for ways to both store ideas and help them interact, users of the idea machine developed methods for archiving, retrieving, and engaging the texts in their keeping.

As the internal logic of the library emerged, a new and latent form of mind did as well. And I don't mean that comparatively or metaphorically: The first forms of artificial intelligence don't go back to Silicon Valley; they go back to Suppiluliuma's neatly organized stacks of cuneiform tablets. But these three traits represent only the start of the particular slant manifested by the idea machine; there are subtler angles at play as well.

Individualism stands out as a fourth trait. While books have always been produced and their use supported by reading communities, such as courts and palaces, synagogues and churches, schools and universities, idiosyncratic readings of those books have always been possible given the private nature of reading. Every book meets an individual brain hemmed by bone and limited by personal experience, which inevitably colors its reception.

A personal reckoning with a text prompts individual acceptance or rejection of its message, and it's a fader, not a toggle: "I understand the message to mean one thing," says one reader; "I understand it to mean another," says his disputant. Sometimes these differing readings spiral off into heresies, other times into new theories of the natural world, and again into multiplying, irreconcilable arguments about human motivations. And as society, particularly Western society, became more individualistic, reading formed a positive feedback loop, both enabling and amplifying this trend.

Fifth, there's secularism. As competing truth claims fueled by individual interpretation came into increased conflict, Western societies looked for ways to negotiate their differences without resorting to violence. Even if there could be no true neutral ground between positions, individuals and groups could ultimately agree to disagree, allowing a truce for the sake of harmony. Toleration and secularism ultimately emerged as the answer.

A lamentable compromise for some and a bona fide improvement for others, secularism meant one group's interpretation of reality didn't supersede another's merely because they claimed the support of divine right or some other unimpeachable authority. This view ultimately paved the way for forms of individual rights, personal freedoms, and the broad-scope cultural liberalism we assume today.

Sixth, we have free speech and open inquiry. If individuals are free to think as they please in a secular society, they're free to say and ask whatever they choose as well. Some of these statements and questions prove scandalous and offensive to many within the society, and they tolerate them only because they feel they must or because the laws and a democratic majority insist they do. Of course, there are gray areas galore, and groups have been ready to tear out hair—their own or their neighbors'—as they fight about crossing whatever arbitrary lines they've drawn.

But most of these expressions of individual interests and pursuits are offered every day with little but procedural fuss among theorists and practitioners of philosophy, law, policy, science, medicine, industry, the arts, and all other sectors of human ingenuity, driving forward advances in all their various fields, usually without fanfare or even public recognition, despite the public's unknowing benefit. Think about medicines and technologies you depend on, the policies to which you appeal, the faith you practice, the art you love, the sports you cheer—all of these have thrived primarily because of the free exchange of ideas and voluntary participation, not coercion.

There's nothing inevitable about the progression I've outlined. It's a fortunate fluke for all I know. But it's inconceivable without the technology of the book. At the start of this volume, I offered historian Barbara Tuchman's observation: "Without books, history is silent, literature dumb, science crippled, thought and speculation at a standstill. Without books, the development of civilization would have been impossible." She went on to call books "engines of change."

We've traced some of that development and change in this book. Books taught us to think; to express ourselves; to domesticate data and organize ideas; to build complex theories and mental models; to painstakingly prove the validity of those theories and models; to negotiate and debate; to relate to those who differ from us over our theories and models, who take offense at them; to relate to each other as equals, even if adversarially; to find ourselves; to advocate for ourselves; to advocate for others.

But to underscore a theme I've emphasized throughout, it wasn't merely the ideas within those books, essential as those were; it was the hardware as well as the software. Ideas don't exist on their own. They are mediated. How we develop and access ideas matters as much as the ideas themselves. The developments described above wouldn't have happened without glyphs on clay, papyrus, parchment, or paper—without words on scrolls, bound in pages, shelved in libraries, and eventually printed by the millions.

And books aren't done with us yet.

What Books Teach Us Now

"A book is . . . a being that speaks on behalf of the dead and acts as an interpreter for the living," said the ninth-century Arab sage Al Jâdiz. At this point, we're familiar with how the book speaks for the dead, but how might the idea machine interpret our present world? I think of two primary challenges facing us:

- Increasing social discord
- The rise of large language models (LLMs)

We've got plenty of other problems, but these two impact many of us daily, and the particular technological bias of books offers, I think, a constructive way to consider both.

Start with social discord. Especially evident in political discourse and cultural conversations around contested values, it seems as though the liberal consensus of individual liberty and free expression is under threat, particularly today on college campuses and social media platforms—but perhaps beyond as well, such as where, for instance, scientific consensus takes on political importance for certain policy initiatives. At the extremes of both Right and Left, mob action overrides debate, and volume replaces evidence.

The history of the book shows that dialogue produces better results than repression, and repression eventually fails anyway. There's a reason the Catholic Church gave up trying to enforce, let alone manage, its *Index of Prohibited Books*. Better to let your Varro sit side by side on the shelf with your Augustine, as Sidonius showed us fifteen centuries ago, or fill your library with every voice imaginable and refute or reconcile them on the page, as Montaigne did five centuries ago.

Outside a few cases of egregious harm, societies thrive on amplifying virtue rather than battling vice, and books themselves remind us that what counted as vice in one generation might count for virtue in the next, and vice versa. Like it or not, cultures evolve, and freeing individuals to think, speak, and live as they choose is the best way to navigate for all parties involved.

What if your minority opinion can't get a fair shake? Look for ways of negotiating or look for alternative avenues to express what matters to you. The temptation to use political force and violence is not only unattractive but also futile and self-defeating. The history of books shows us that much; it also offers more fruitful models for engagement.

And what of generative AI and LLMs? Many in my circle live and work in and around the humanities and education more broadly. I'd say the majority are either concerned about AI or actively opposed. But if you see books as part of the human endeavor to domesticate the world we inhabit—to preserve information and expand access to it, to organize and utilize data and analysis, to build on what we know and broaden what we can learn—then it's possible to regard LLMs as part of the idea machine, as an extension of its affordances. LLMs stand in a line running from the libraries of Suppiluliuma and Ashurbanipal to Aristotle's collection and the archivists and catalogers at Alexandria, to Cicero, Origen, and Eusebius, to Hernando Colón and Conrad Gessner, to Paul Otlet, Vannevar Bush, and J. C. R. Licklider.

Of course, identifying continuities doesn't rule out discontinuities; as with the disruptions caused by the advent of print and the subsequent flood of publication, we'll continue to experience upheavals as we negotiate new norms and procedures. But even then, the history of the book encourages optimism, not pessimism. If it didn't, we'd find ourselves back in the debate between Socrates and Phaedrus, Theuth and Thamus. Socrates may have won the argument, but we've done just fine ignoring the conclusion. Better to follow his example than his pronouncements on this point.

As we humans use our tools for what we, as humans, value, we will find humane applications that further our interests and expand our native capabilities, despite the inevitable disruption of technological change. How will that ultimately affect education, scientific research, literary analysis, artistic production, and other areas of endeavor? Participation is the only way to have a say in the answer. The idea machine built our world and still shapes our future. Its users determine how.

THANKS

"Nobody's fault but mine," sang bluesman Blind Willie Johnson. Authors can say much the same about their books, but several folks aided and abetted me in writing this one.

Agent Andrew Wolgemuth believed in the project from the start. So did Jeff Goins of Fresh Complaint. Both helped me refine the argument and encouraged my keystrokes along the way. Jake Bonar led the charge at Prometheus. Patricia Stevenson and Jen Kelland improved my prose and saw the book through production, while Alyssa Griffin tackled the marketing. My friend Mike Burns contributed the graphs for the x-, y-, and z-axes.

Several friends read the manuscript in part or in full or heard sketches of the argument at various stages: Jeremy Lott, Adam Hill, Larry Stone, Larry Wilson, Hannah Williamson, Stuart Kells, Erik Rostad, Dave Schroeder, Hollis Robbins, and Henry Oliver.

Special thanks to the Full Focus family, Tom Singleton, the Boys of Barbecue, and all the readers of MillersBookReview.com. You helped make this happen.

Very special thanks to my favorite human, Megan, for not only encouraging me through the many years of work but also making it possible to finish. And to my kids—Fionn, Felicity, Moses, Jonah, and Naomi—for continually supplying me with hope for the future (and quite a few laughs).

If I could have done it without you all, I wouldn't have wanted to.

NOTES

In all but one instance, I've omitted URLs from the citations below. Readers can locate the referenced materials online using names, titles, keywords, or a combination thereof. I've used the Loeb Classical Library wherever convenient, citing volumes with the abbreviation LCL followed by the volume number. After a particular edition of a book has been mentioned, I simplify subsequent references. Unless otherwise noted, I've taken biblical quotations from the New Revised Standard Version. For the sake of consistency, I've standardized the Roman poet's name as Vergil throughout the book. (Apologies to all who prefer Virgil. Half of us were bound to be disappointed by something; it might as well be that.)

Introduction

1   "Whenever I want": Bill Gates, "10 of Bill Gates's Favorite Books about Technology," *MIT Technology Review*, February 27, 2019.

2   "Tolle lege" / "Take and read": Augustine, *Confessions* 8.12.29.

2   "The writer has": Eugene Vodolazkin, "In Memory of Sharov," trans. Google Translate, *Izvestia*, August 18, 2018.

2   "It's so strange": Eugene Vodolazkin, *The Aviator*, trans. Lisa C. Hayden (Oneworld, 2018), 127.

4   "human chain": Rodrigo Fresán and Rodrigo Rey Rosa, "Were Libraries Borges's Universe or the Other Way Around?" *Literary Hub*, May 31, 2019.

4   "I placed my finger or some": Augustine, *Confessions* 8.12.30, trans. Philip Burton (Everyman's Library, 2001). Alan Jacobs notes this detail in "Christianity and the Future of the Book," *The New Atlantis*, Fall 2011.

4   Pagan and Christian authors: Sidonius Apollinaris, *Letter* 2.9.4 (to Donidius), LCL 296.

5   "Books are a uniquely portable magic": Stephen King, *On Writing* (Simon & Schuster, 2002), 96.

5   "Any sufficiently advanced": Jeff Prucher, ed., *Brave New Words* (Oxford University Press, 2007), 22.

7   "I am a product": C. S. Lewis, *Surprised by Joy* (Harcourt, 1955), 8.

7   "When books are bequeathed": Georgios Boudalis, *The Codex and Crafts in Antiquity* (Bard, 2018), 2.

10   By the time Herodotus was writing: Herodotus, *Histories* 5.58.3.

10   Demosthenes said a small sheet: Demosthenes, *Against Dionysodorus* 1.

10   Papyrus wasn't exactly cheap: William V. Harris, *Ancient Literacy* (Harvard University Press, 1989), 95.

Marginalia: Ideas across Time

11   Simple, two-dimensional grid: We developed the idea grid at Full Focus to discuss communication with coaching clients. See Michael Hyatt, *No-Fail Communication* (Michael Hyatt & Co., 2020), 17; Michael Hyatt, *The Vision-Driven Leader* (Baker, 2020), 87. I added the z-axis for use in this analysis.

Chapter 1: Socrates, Technophobe?

17   "phrase-collector" / "beetroot, light phrases": Aristophanes, *The Frogs*, trans. Gilbert Murray, in *7 Famous Greek Plays*, ed. Whitney J. Oates and Eugene O'Neill Jr. (Vintage, 1950), 395, 401. For "chatter-juice," see Alan H. Sommerstein's translation.

18   "Come, no more line-for-lines!" / "You traitor" / "Oft from long campaigns": Aristophanes, *The Frogs*, 426, 430, 411.

19   Socrates discussed the subject: Xenophon, *Memorabilia* 4.2.1–39, trans. Amy L. Bonnette (Cornell University Press, 1994).

21   "Eventually he [Phaedrus] borrowed" / "I suspect": Plato, *Phaedrus*, trans. Robin Waterfield (Oxford University Press, 2009), 4, 5.

23   "Your Highness" / "One person has" and subsequent statements from Thamus: Plato, *Phaedrus*, 68, 69.

24   "Once any account" / "always needs its father" / "They just go on": Plato, *Phaedrus*, 70.

25   "Written words can do" / "Quite so": Plato, *Phaedrus*, 70.

25   "clear and reliable": Plato, *Phaedrus*, 69.

25   "Reading collectively": Xenophon, *Memorabilia* 1.6.14.

26   "orders and is the cause": Plato, *Phaedo*, trans. F. J. Church (Liberal Arts Press, 1954), 50–51.

27   "Read out from the scroll": Plato, *Phaedrus*, 24.

27   Recourse to quotably precise: David R. Olson, "Why Literacy Matters, Then and Now," in *Ancient Literacies*, ed. William A. Johnson and Holt N. Parker (Oxford University Press, 2011), 385–403.

Marginalia: The Genesis of Writing

29   Recognizable writing appeared: Andrew Robinson, "Writing Systems," in *Oxford Companion to the Book*, ed. Michael F. Suarez and H. R. Woudhuysen (Oxford University Press, 2010), 1:1–10.

29   Writing became a political: Henri-Jean Martin, *History and Power of Writing*, trans. Lydia G. Cochrane (University of Chicago Press, 1994), 13–15.

29   "concealed . . . in their underwear": Mathilde Touillon-Ricci, "Trade and Contraband in Ancient Assyria," British Museum, April 2, 2018. On this and other smuggling schemes, see Gojko Barjamovic, *A Historical Geography of Anatolia in the Old Assyrian Colony Period* (Museum Tusculanum Press, 2011), section 4.9.

Chapter 2: Upgrading the Mind

31   "there appears to be": Egert Pöhlmann and Martin L. West, "The Oldest Greek Papyrus and Writing Tablets Fifth-Century Documents from the 'Tomb of the Musician' in Attica," *Zeitschrift für Papyrologie und Epigraphik* 180 (2012): 4.

33   "The magic of writing": David R. Olson, *The World on Paper* (Cambridge University Press, 1994), xv. Olson extends this line of thought throughout this book and also in his follow-up, *The Mind on Paper* (Cambridge University Press, 2016).

33   Without that extended circuit: Andy Clark, *Supersizing the Mind* (Oxford University Press, 2011), xxv–xxvi; Andy Clark, *The Experience Machine* (Pantheon, 2023), 147–180; Andy Clark, "Magic Words," in *Language and Thought*, ed. Peter Carruthers and Jill Boucher (Cambridge University Press, 1998).

33   "He fashioned his poem": Suetonius, *Life of Vergil* 22, LCL 38. Vergil used this licking analogy in the *Aeneid* itself. In book 8, as John Dryden translated it, the wolf's "fawning tongue . . . lick'd their [Remus and Romulus's] tender limbs, and form'd them as they fed."

34   "After writing a first": Suetonius, *Life of Vergil* 23–24.

34   "She proceeds to set down": Ovid, *Metamorphoses* 9.521ff., LCL 43.

36   "We must first criticise": Quintilian, *Institutio Oratoria* 10.3.5, 6, 7, LCL 127.

36   "Murder your darlings": Arthur Quiller-Couch, *On the Art of Writing* (Putnam, 1916), 281.

37   "We can see he made": Merrit Kennedy, "Ancient Greek Scroll's Hidden Contents Revealed Through Infrared Imaging," NPR, October 4, 2019.

37   "Sometimes . . . the most admirable": Quintilian, *Institutio Oratoria* 10.3.33.

39   It's difficult to imagine: See Walter J. Ong's lengthy discussion in *Orality and Literacy*, 30th anniv. ed. (Routledge, 2012), 31–76; for a concrete example, see Christina Thompson's description of Polynesian oral culture in *Sea People* (Harper, 2019), 126–138.

39   "Something ineffable" / "a kind of shifting": Lynn Hunt, "How Writing Leads to Thinking," *Perspectives on History* 48, no. 2 (February 1, 2010).

40   Functional magnetic resonance imaging: Anna Abraham, *The Neuroscience of Creativity* (Cambridge University Press, 2018), 219, 221.

41   "Plato did not cease": Dionysius of Halicarnassus, *On Literary Composition* 25, trans. W. Rhys Roberts (Macmillan, 1910). Quintilian tells the same story in *Institutio Oratoria* 8.6.64.

Marginalia: Notes to Self

42   "The most learned": This and subsequent quotes from Sidonius Apollinaris, *Letter* 5.2, "To His Friend Nymphidius," in *The Letters of Sidonius*, trans. O. M. Dalton, Vol. 2 (Oxford University Press, 1915).

43   "artificial memory": Earle Havens, *Commonplace Books* (Beinecke, 2001), 14.

43   "We ought in some": Havens, *Commonplace Books*, 16.

Chapter 3: Tools for Thinking

45   "My husband is dead" / "No doubt": Edgar H. Sturtevant, "The Hittite Tablets from Boghaz Kevi," *The Classical Weekly* 18, no. 22 (April 20, 1925). See also Eric H. Cline, *1177 B.C.*, rev. ed. (Princeton University Press, 2021), 64–66.

45   "My father asked" / "In the old days": Theo P. J. van den Hout, "Miles of Clay: Information Management in the Ancient Near Eastern Hittite Empire," *Fathom Archive*, October 26, 2002.

47   "From this and many": Van den Hout, "Miles of Clay."

48   When archaeologists unearthed: D. T. Potts, "Before Alexandria," in *The Library of Alexandria*, ed. Roy McLeod (I. B. Tauris, 2004), 23.

48   "psychopathic bookworm": Mark Brown, "British Museum Shines Light on Assyrian 'King of the World,'" *Guardian*, June 19, 2018.

48   "I wrote on tablets": Lionel Casson, *Libraries in the Ancient World* (Yale University Press, 2001), 10.

48   "Collect all the tablets": Jeanette C. Finke, "The Babylonian Texts of Nineveh," *Archiv für Orientforschung* 50 (2003/2004).

48   "was to provide large": Eleanor Robson, "Reading the Libraries of Assyria and Babylonia," in *Ancient Libraries*, ed. Jason König et al. (Cambridge University Press, 2013), 45.

50   "Once an idea": Tyler Cowen, *Stubborn Attachments* (Stripe, 2018), 97.

50   "A book is a machine": I. A. Richards, *Principles of Literary Criticism* (Routledge Classics, 2001), vii.

50   "Some authorities": Diogenes Laertius, *Lives of Eminent Philosophers* 3.9, LCL 184.

50   On at least one occasion: Diogenes Laertius, *Lives* 2.81.

50   Plato borrowed ideas: Diogenes Laertius, *Lives* 3.9–18.

50   One comprehensive study: Konstantinos Sp. Staikos, *Books and Ideas: The Library of Plato and the Academy*, trans. Nikos Koutras (Oak Knoll, 2013), 227–231.

51   Based on a reconstruction: Staikos, *Books and Ideas*, especially 165–209.

51   "research engine": Simon Critchley, "Athens in Pieces," *New York Times*, February 6, 2019.

51 A generation of Plato's death: Staikos, *Books and Ideas*, 161–162.
51 Aristotle "the Mind" / "The lecture room": Yasuhira Y. Kanayama, "Plato's Wax Tablet," in *Soul and Mind in Greek Thought*, ed. Marcelo D. Boeri et al. (Springer, 2019), 94.
52 Plato compared him: Diogenes Laertius, *Lives* 5.1.2.
52 Aristotle's library: "Archeologists Believe They Have Found Aristotle's School," Associated Press, January 15, 1997.
52 "We should select": Aristotle, *Topics* 1.14, in *The Basic Works of Aristotle*, ed. Richard McKeon (Modern Library, 2001).
52 "there exists much": Aristotle, *On Sophistical Refutations* 34.184, in *The Basic Works of Aristotle*.
52 The same was true: See Armand Marie Leroi, *The Lagoon* (Penguin, 2014).
53 "were therefore able": Strabo, *Geography* 13.1.54, LCL 223.
54 The Hattusa catalog: Casson, *Libraries*, 5–7.
54 To manage the stampede: Finke, "Babylonian Texts."
55 "Greek Big Data": Dennis Duncan, *Index, A History of the* (Allen Lane, 2021), 35.

Marginalia: Cicero's Analog AI

56 He supposedly possessed: William A. Johnson, "Cicero and Tyrannio: '*Mens Addita Videtur Meis Aedibus*' ('*Ad Atticum*' 4.8.2)," *The Classical World* 105, no. 4 (2012), 473.
56 "better able to philosophize": Strabo, *Geography* 13.1.54.
56 "It is said that after the library": Plutarch, *Sulla* 26, LCL 80.
57 "You will be surprised" / "Your men have beautified" / "Since Tyrannio has arranged": Cicero, *Letter* 4.4a, 4.5, 4.8, LCL 7.
57 "recovered its intelligence": George W. Houston, *Inside Roman Libraries* (University of North Carolina Press, 2014), 219.
57 Duke University classicist: Johnson, "Cicero and Tyrannio," 472. Incidentally, Cicero's use of the Latin *mens* here mirrors Socrates's use of the Greek *nous* when complaining about Anaxagoras's materialism.

Chapter 4: Clarity at Scale—Almost

59 "I have found": This and subsequent quotes taken from 2 Kings 22 and 23.
59 No one can identify: Probably a stretch of text known as the Deuteronomic Code, chapters 12–26, in the final edition of Deuteronomy.
60 "This was one": William Schniedewind, *How the Bible Became a Book* (Cambridge University Press, 2004), 91.
61 Standing on a dais / "They gave the sense": Nehemiah 8.3, 8.8.
62 Moses . . . "required them": Philo, *Hypothetica* 7.12–14, LCL 363.
62 "all that Moses committed": Juvenal, *Satires* 14, LCL 91.
62 "Any one of them": Philo, *Hypothetica* 7.12–14.

62   "On the day called": Justin Martyr, *First Apology* 67, in *Ante-Nicene Fathers*, ed. Alexander Roberts and James Donaldson, Vol. 1 (Hendrickson, 1994).

63   "different Hebrews" / "several distinct dialects": James L. Kugel and Rowan A. Greer, *Early Biblical Interpretation* (Westminster Press, 1986), 29.

63   "I haven't the faintest": Hannah Furness, "Keep It Simple: Even I Struggle with Shakespeare, Says Hytner," *London Telegraph*, October 10, 2013.

63   Linguists such as John McWhorter: John McWhorter, *Words on the Move* (Henry Holt, 2016), 56–59, 75, 85–94, 98, 115, 193.

63   "false friends": Kugel and Greer, *Early Biblical Interpretation*, 29.

63   Translation eventually filled: Jaroslav Pelikan, *Whose Bible Is It?* (Penguin, 2005), 59–60.

63   "salvation written in their hearts": Irenaeus, *Against Heresies* 3.4.2, trans. Dominic J. Unger and Irenaeus M. C. Steenberg (Newman Press, 2012).

64   "in more need of restraints": Philostorgius, *Ecclesiastical History* 2.5., trans. Edward Walford, in *The Ecclesiastical History of Sozomen* (Henry G. Bohn, 1855).

64   "I have believed your scriptures": Augustine, *Confessions* 12.10.10, trans. Maria Boulding (New City Press, 1997).

64   "hard to understand" / "unschooled and unstable": 2 Peter 3.16, my paraphrase.

64   "the abomination" / "Let the reader": See Matthew 24.15 and Mark 13.14, English Standard Version.

65   "It would all have been": Henry Chadwick, *The Early Church*, rev. ed. (Penguin, 1993), 33.

65   "were handed on": Luke 1.2.

66   True for the Gospel of Mark: Eusebius, *The History of the Church* 2.15, trans. G. A. Williamson (Penguin, 1965).

66   "to preach with sound": Titus 1.9.

66   "appointed those we": 1 Clement 44.2, LCL 24.

66   "To my dearest lady sister": AnneMarie Luijendijk, *Greetings in the Lord* (Harvard Theological Studies, 2008), 71.

66   Irenaeus used his position: Irenaeus, *Against Heresies* 1.31.1, trans. Dominic J. Unger and John J. Dillon (Newman Press, 1992).

66   "We, my brothers": Eusebius, *The History of the Church* 6.12.

68   "ecclesiastical tradition" / "the succession": Eusebius, *Church History* 3.25, in *Nicene and Post-Nicene Fathers*, Vol. 1, second series, ed. Philip Schaff et al. (Hendrickson, 1994).

68   Bishops oversaw the copying: See, for instance, Luijendijk, *Greetings in the Lord*, 144–151; Anthony Grafton and Megan Williams, *Christianity and the Transformation of the Book* (Belknap, 2006).

68   The institutional church: See Elaine Pagels, *The Gnostic Gospels* (Random House, 1979), esp. chaps. 2, 5.

Marginalia: By the Book

69   "worthy soul" / "If you desire": Dante Alighieri, *Paradise*, Canto 5, in *The Portable Dante*, trans. Mark Musa (Penguin, 1995).

69   "Caesar I was": Dante, *Paradise*, Canto 6.

69   "Whoever is convicted" / "Slaves caught" / "If anyone sings": Allan Chester Johnson et al., *Ancient Roman Statutes* (University of Texas Press, 1961), 11–12, 15.

70   The first step was: Lorena Atzeri, "Roman Law and Reception," *Europäische Geschichte Online*, November 20, 2017; Peter Sarris, *Justinian* (Basic, 2023), 122–132, 422–424.

70   Free-floating records: Sarris, *Justinian*, 128, 453.

70   "Thus . . . the entire": Sarris, *Justinian*, 131.

Chapter 5: No Books for You

73   "He is a wretched" / "I have sent back": Cicero, *Letter* 2.20, 22.

73   Copying books was tedious: Houston, *Inside Roman Libraries*, 13–22.

75   As much as a week's pay: Rex Winsbury, *The Roman Book* (Duckworth, 2009), 19–20; Harris, *Ancient Literacy*, 194–195.

75   Pliny the Elder tells of a tiff: Pliny, *Natural History* 13.21, LCL 370.

76   Some doubt the story: C. H. Roberts and T. C. Skeat, *The Birth of the Codex* (Oxford University Press, 1987), 6.

76   "the material on which": Pliny, *Natural History* 13.21.

76   "river of letters" / "monolith of characters": Winsbury, *Roman Book*, 35; Harry Y. Gamble, *Books and Readers in the Early Church* (Yale University Press, 1995), 48.

77   "Tell me, O Muse": Homer, *Odyssey*, book 1, lines 1–10, LCL 104.

77   Without spaces: Winsbury, *Roman Book*, 35–44; Paul Saenger, *Space Between Words* (Stanford University Press, 1997), 6–8, 11–12, 39, 122; William A. Johnson, *Readers and Reading Culture in the High Roman Empire* (Oxford University Press, 2010), 17–31.

78   "Ignorant schoolboys": Aulus Gellius, *Attic Nights* 13.31.1, LCL 200.

78   Literacy is near universal: "Literacy Rates Continue to Rise from One Generation to the Next," Fact Sheet No. 45, UNESCO Institute for Statistics, September 2017.

78   Even with Rome's extensive: Harris, *Ancient Literacy*, 141, 272.

78   "The ancient world": Saenger, *Space Between Words*, 11.

79   "were designed for clarity": Johnson, *Readers and Reading Culture*, 20.

79   "works written expressly": Harris, *Ancient Literacy*, 227.

79   Tools for self-advancement: According to Spanish essayist Irene Vallejo, "Aristocrats, proud of their cultural superiority, had no interest in making things easier for readers they saw as upstarts—that is, those with less access to education—so that they could sneak into the exclusive fiefdom of books." Irene Vallejo, *Papyrus*, trans. Charlotte Whittle (Knopf, 2022), 336.

79 Slaves were plentiful: Mark Cartwright, "Slavery in the Roman World," *World History Encyclopedia*, November 1, 2013; Perry Anderson, *Passages from Antiquity to Feudalism* (Verso, 2013), 38.

80 "writings of the ancients": Houston, *Inside Roman Libraries*, 13.

81 "Now, as you love" / "This gift depends": Cicero, *Letter* 1.20, 2.1.

82 "We cannot be sure": Houston, *Inside Roman Libraries*, 31.

82 "Seek out Secundus" / "with its doorposts": Martial, *Epigrams* 1.2, 1.117, LCL 94.

82 "I am much obliged": Cicero, *Letter* 2.4.

83 "Thyillus asks you": Cicero, *Letter* 1.9.

83 "He would frequently" / "No book was so bad": Pliny the Younger, *Letters* 3.5 (to Baebius Macer), LCL 55.

83 "We also . . . ought to" / "Reading . . . enable[s]": Seneca, *Epistulae Morales* 84, LCL 76.

84 "in a small hand": Pliny the Younger, *Letters* 3.5.

84 Julius Caesar: Roberts and Skeat, *Birth of the Codex*, 19.

84 "designed for dipping into": Matthew Nicholls, "Parchment Codices in a New Text of Galen," *Greece and Rome*, second series, 57, no. 2 (October 2010).

85 Galen to ably serve: Nicholls, "Parchment Codices in a New Text of Galen."

85 "You, who wish": Martial, *Epigrams* 1.2.

85 "little book" being "thumbed everywhere": Martial, *Epigrams* 8.1, 3, LCL 95.

85 Statistical analysis: Larry W. Hurtado, *The Earliest Christian Artifacts* (Eerdmans, 2006), 53.

85 One scholar compared: Eric G. Turner, *The Typology of the Early Codex* (Wipf & Stock, 2011), 1.

Marginalia: St. Augustine's Tome

86 "I decided to write": Augustine, *Retractions* 69, trans. Mary Inez Bogan (Catholic University Press of America, 1968).

86 In 409, a priest asked: Gerard O'Daly, *Augustine's* City of God, 2nd ed. (Oxford University Press, 2020), 32.

86 Other aspects of *The City*: O'Daly, *Augustine's* City, 66–67.

87 Augustine buttonhole Vergil: For a detailed look at the authors Augustine cited or referenced, see O'Daly, *Augustine's* City, 266–297.

87 "once an idea": Cowen, *Stubborn Attachments*, 97.

87 "The Rome he deconstructs": Gillian Clark, "City of Books," in *The Early Christian Book*, ed. William E. Klingshirn and Linda Safran (Catholic University Press of America, 2007), 125.

Chapter 6: Turning a New Page

91 The cities of Mediterranean Europe: Travel times listed here come from Stanford's Orbis project (orbis.stanford.edu).

91 "holy internet": Michael B. Thompson, "The Holy Internet," in *The Gospels for All Christians*, ed. Richard Bauckham (Eerdmans, 1998).

91 "And when this letter": Colossians 4.16.

92 "His letters are weighty": 2 Corinthians 10.10.

92 Author of the Second Epistle of Peter: 2 Peter 3.15–16.

92 Approached by an elderly woman: *Shepherd of Hermas*, Vision, 2.1, 2, 4, LCL 25.

92 "soon after . . . we find papyrus": Kim Haines-Eitzen, *Guardians of Letters* (Oxford University Press, 2000), 77.

94 "When you have heard" / "This account Gaius": *Martyrdom of Polycarp* 20.1–2, 23, LCL 24; Haines-Eitzen, *Guardians*, 79–80.

94 "because . . . I have not": Ignatius, *Letter to Polycarp* 8, LCL 24.

94 "We have forwarded": Polycarp, *To the Philippians* 13, LCL 24.

94 Time between the request: Gamble, *Books and Readers*, 110.

94 "In all this . . . we glimpse": Gamble, *Books and Readers*, 112.

95 "Many have undertaken": Luke 1.1–2.

95 "Whenever someone arrived": Papias, fragment 3, LCL 25. See also Eusebius, *History of the Church*, 3.39.

95 "I too decided": Luke 1.3–4.

95 "Now Jesus did many": John 20.30–31.

96 "There are also": John 21.25.

96 T. C. Skeat, keeper of manuscripts: T. C. Skeat, "The Origin of the Christian Codex," in *The Collected Biblical Writings of T. C. Skeat*, ed. J. K. Elliott (Brill, 2004).

96 Other scholars have argued: See Gamble, *Books and Readers*; Hurtado, *Earliest Christian Artifacts*; E. Randolph Richards, *Paul and First-Century Letter Writing* (IVP, 2004).

96 One third-century Egyptian: Haines-Eitzen, *Guardians*, 96–97.

98 "The impetus among early": Larry W. Hurtado, *Destroyer of the Gods* (Baylor University Press, 2016), 137.

98 "Do your best to come": 2 Timothy 4.9, 13.

98 Compiling these *testimonia*: Gamble, *Books and Readers*, 65.

98 The volumes Paul requested: See Richards, *Paul and First-Century Letter Writing*, 210–223.

98 "Christian texts came to be inscribed": Gamble, *Books and Readers*, 66.

98 The story of Peregrinus: Lucian, *Death of Peregrinus* 11, 12, LCL 302.

99 Population growth figures: Based on the numbers in Rodney Stark, *Rise of Christianity* (Harper, 1997), 6–7.

99 Peregrinus's jail date: Gilbert Bagnani, "Peregrinus Proteus and the Christians," *Historia* 4, no. 1 (1955).

99 Previous critics of Christianity: Robert Louis Wilken, *The Christians as the Romans Saw Them* (Yale University Press, 2003), 108–109.

99 "What are the things": *The Passion of the Scillitan Martyrs*, in *Ante-Nicene Fathers*, ed. Allan Menzies, Vol. 9 (Hendrickson, 1994).

99 Gamble notes: Gamble, *Books and Readers*, 299.

100   "An imperial decree": Eusebius, *History of the Church* 8.2.4.

100   "The gates having been": Lactantius, *Of the Manner in Which the Persecutors Died* 12, in *Ante-Nicene Fathers*, ed. Alexander Roberts et al., Vol. 7 (Hendrickson, 1994).

100   Books destroyed at Cirta: *Gesta Apud Zenophilum*, taken from two translations: O. R. Vassals-Phillips, *The Work of St. Optatus* (Longmans, 1917); Mark J. Edwards, *Optatus: Against the Donatists* (Liverpool, 1997).

101   Constantine's book order: Eusebius, *Life of Constantine* 4.36, 37, in *Nicene and Post-Nicene Fathers*, Vol. 1, second series, ed. Philip Schaff et al. (Hendrickson, 1994).

101   "Christianity spread from": Mary Beard, *SPQR* (Liveright, 2015), 520, emphasis added.

101   "As it spread through": Martin, *History and Power of Writing*, 110.

102   In the second century, Christian books: Hurtado, *Earliest Christian Artifacts*, 48–49, 53, 91, 93.

103   "The book is a letter": Florence Dupont, "The Corrupted Boy and the Crowned Poet," in *Ancient Literacies*, 158.

Marginalia: Reimagining the Book

104   "There are . . . twenty-two books": Gamble, *Books and Readers*, 134.

104   Asks Firmus about distributing: Gamble, *Books and Readers*, 135.

105   New benefits: List adapted from Gamble, *Books and Readers*, 49–66; Jacobs, "Christianity and the Future of the Book."

105   Not only were codices easy: Gamble, *Books and Readers*, 55.

105   "the idea . . . that wisdom": James O'Donnell, *Augustine* (Harper, 2006), 328.

Chapter 7: What the Monks Did

107   Anthony couldn't shake: Athanasius, *Life of St. Anthony* 1–3, in *Nicene and Post-Nicene Fathers*, Vol. 4, second series, ed. Philip Schaff et al. (Hendrickson, 1994).

107   "holy bands of men": Athanasius, *Life of St. Anthony*, 44.

109   "If he is illiterate": *Rules of Saint Pachomius* 139–140, in *Pachomian Koinonia*, Vol. 2, trans. Armand Veilleux (Cistercian Publications, 1981).

109   "Every western aristocrat": Chris Wickham, *The Inheritance of Rome* (Penguin, 2010), 29–30.

109   Wealthy gained familiarity: Peter Brown, *Through the Eye of a Needle* (Princeton University Press, 2014), 164.

110   "In each house" / "They read": Philo of Alexandria, *On the Contemplative Life* 25, 28–29, LCL 363.

110   "Always have a book": Jerome, *Letter* 125 (to Rusticus), LCL 262.

110   "Being very learned" / "She laboriously went" / "Wherefore also": Palladius, *Lausiac History* 55.3, Internet Medieval Source Book, Fordham University.

111   The thought lodged: N.185/5.38, in *The Anonymous Sayings of the Desert Fathers*, trans. John Wortley (Cambridge University Press, 2013).

111   "The conscience is nature's book": Mark the Monk, *On the Spiritual Law* 187, in *Counsels on the Spiritual Life*, trans. Tim Vivian (St. Vladimir's Seminary Press, 2009).

111   Dorotheos of Gaza compared: Dorotheos of Gaza, *On Traveling the Way of God with Vigilance and Sobriety*, in *Discourses and Sayings*, trans. Eric P. Wheeler (Cistercian, 2008), 164.

112   "I have spent twenty years": N.541, in *Anonymous Sayings*.

112   "Words are not wiser: Mark the Monk, *Concerning Those Who Imagine That They Are Justified by Works* 50, in *Counsels on the Spiritual Life*.

112   Story of Abba Serapion: N.566/15.117, in *Anonymous Sayings*.

112   "The prophets made the books": N.228/10.191, in *Anonymous Sayings*.

113   "The days will come": N.758, in *Anonymous Sayings*.

113   "By the middle" / "a byproduct": Haines-Eitzen, *Guardians*, 38–39.

114   Jerome . . . instructed his prospective monk: Jerome, *Letter* 125.

114   "Christianity has no sacred tongue": Robert Louis Wilken, *The First Thousand Years* (Yale University Press, 2012), 2.

114   "pleases me most": Cassiodorus, *Institutions of Divine and Secular Learning* 1.30.1, trans. James W. Halporn (Liverpool University Press, 2004).

116   "In adding detail": Anita Radini et al., "Medieval Women's Early Involvement in Manuscript Production Suggested by Lapis Lazuli Identification in Dental Calculus," *Science Advances* 5, no. 1 (January 9, 2019).

116   In his two-volume study: George Haven Putnam, *Books and Their Makers in the Middle Ages* (Hillary House, 1962), 1:52–55.

116   Examples have only multiplied: See, for example, Mary Wellesley, *The Gilded Page* (Basic, 2021), 154–159.

117   "The obliteration of classical texts": Andrew Pettegree and Arthur der Weduwen, *The Library* (Basic, 2021), 36.

118   Hence Augustine's surprise: Augustine, *Confessions* 6.3.3.

118   When Philip finds the Ethiopian eunuch: Acts 8.30, emphasis added.

118   When Cyril of Jerusalem directs: Cyril of Jerusalem, *Protocatechesis* 14, in *Nicene and Post-Nicene Fathers*, Vol. 7, second series, ed. Philip Schaff et al. (Hendrickson, 1994).

118   Appointed readers: Saenger, *Space Between Words*, 6–9, 72.

119   "When proper words make": Augustine, *On Christian Doctrine* 3.2.2–3, in *Nicene and Post-Nicene Fathers*, Vol. 2, first series, ed. Philip Schaff (Hendrickson, 1994).

119   Irenaeus said the Gnostics: Irenaeus, *Against Heresies* 3.7.2.

119   To ensure correct readings: Hurtado, *Earliest Christian Artifacts*, 178–179.

121   Widespread adoption: Saenger, *Space Between Words*, 52–82.

121   Earliest Greek manuscripts: Saenger, *Space Between Words*, 10, 55, 59, 70–71, 93.

Marginalia: Upping the Data Game

122   "I am the first": Eusebius, *History of the Church* 1.1.
122   Historians, who tended to fabricate: Andrew Louth, introduction to Eusebius, *History of the Church* (Penguin, 1989), xx.
122   "I have . . . picked out": Eusebius, *History of the Church* 1.1.
122   "girls trained": Eusebius, *History of the Church* 6.23.
122   End result was a research library: Paul Hartog, "Pamphilus the Librarian and the Institutional Legacy of Origen's Library in Caesarea," *Theological Librarianship* 14, no. 1 (2021).
123   Unique technological affordances: Grafton and Williams, *Christianity and the Transformation of the Book*, 131.
123   It took forty volumes: Grafton and Williams, *Christianity and the Transformation of the Book*, 88, 90, 105.
123   Jerome used it: Grafton and Williams, *Christianity and the Transformation of the Book*, 236.
123   The work proved so helpful: Grafton and Williams, *Christianity and the Transformation of the Book*, 136–140.
124   By numbering the discrete sections: Grafton and Williams, *Christianity and the Transformation of the Book*, 199.
124   "the world's first hot links" / "No early creator": Grafton and Williams, *Christianity and the Transformation of the Book*, 199–200. See also Jeremiah Coogan, *Eusebius the Evangelist* (Oxford University Press, 2023), 28–58.

Chapter 8: God's Mom and Books for All

126   Ambrose, fourth-century bishop: Luigi Gambero, *Mary and the Fathers of the Church*, trans. Thomas Buffer (Ignatius, 1999), 192.
126   Ambrose was one of the first: But not *the* first. Origen beat him to the punch. "Mary knew the Law," he wrote; "she was holy, and had learned the writings of the prophets by meditating on them daily." Origen, *Homilies on Luke* 6.7, trans. Joseph T. Lienhard (Catholic University of America Press, 2009). I'm indebted to Zena Hitz for pointing this out.
126   "Indeed . . . how could she": Gambero, *Mary and the Fathers*, 200–201.
126   Miles highlights other writers: Laura Saetveit Miles, "The Origins and Development of the Virgin Mary's Book at the Annunciation," *Speculum* 89, no. 3 (July 2014).
126   "She was certain": Bede, *Homily* 1.3, in *Homilies on the Gospels: Book One, Advent to Lent*, trans. Lawrence T. Martin and David Hurst (Cistercian, 1991).
127   Tour of the Italian peninsula: Johannes Fried, *Charlemagne*, trans. Peter Lewis (Harvard University Press, 2016), 207.
127   Renewed armies: Fried, *Charlemagne*, 259, 373ff.
127   Literacy had always been: F. Donald Logan, *The History of the Church in the Middle Ages* (Routledge, 2002), 59–63.

128   "We exhort you . . . to study": Charlemagne, Letter to Baugaulf of Fulda, c. 780–800, Internet Medieval Source Book, Fordham University.

129   While covering many subjects: John J. Contreni, "Learning for God," *Journal of Medieval Latin* 24 (2014).

129   Majority of ancient Latin manuscripts: Bernhard Bischoff, *Manuscripts and Libraries in the Age of Charlemagne*, trans. Michael Gorman (Cambridge University Press, 2007), 99.

129   "Carolingian scribes were": Steven Roger Fischer, *A History of Reading*, rev. ed. (Reaktion, 2019), 166.

129   In the 250 years between: Contreni, "Learning for God."

130   "he found her praising God": Miles, "Origins and Development."

130   Otfrid likely mined: Miles, "Origins and Development."

131   "In the Brunswick Casket": Miles, "Origins and Development."

131   Mary was pictured: David Lyle Jeffrey, *People of the Book* (Eerdmans, 1996), 215.

131   Encouraging education for ecclesiastics: Chris Wickham, *Medieval Europe* (Yale University Press, 2017), 72.

131   "The Carolingian laity": Rosamond McKitterick, *The Carolingians and the Written Word* (Cambridge University Press, 1990), 270, 272.

133   Before Vikings had ravaged: King Alfred's preface to the translation of Gregory's *Pastoral Care*, in *Documents Illustrating Early Education in Worcester*, ed. Arthur F. Leach (Worcestershire Historical Society, 1913), 6.

133   Alfred conceived a plan: Marc Morris, *The Anglo-Saxons* (Hutchinson, 2021), 231–238.

133   "By the twelfth century": Nicholas Orme, *Medieval Children* (Yale University Press, 2001), 238–239.

133   Fees were required to attend: Seb Falk, *The Light Ages* (Norton, 2021), 31.

133   "In an attempt to drive": Falk, *Light Ages*, 82.

134   "torrent of translations": Falk, *Light Ages*, 86.

134   Queen St. Radegund founded: Jane Tibbetts Schulenberg, *Forgetful of Their Sex* (University of Chicago Press, 2001), 96; Nicholas Everett, "Literacy from Late Antiquity to the Early Middle Ages, c. 300–800 AD," in *The Cambridge Handbook of Literacy*, ed. David R. Olson and Nancy Torrance (Cambridge University Press, 2009), 371.

134   Evidence for female learning: Contreni, "Learning for God," 122. See also Everett, "Literacy from Late Antiquity," 370–371. "Such evidence," says Everett, "dwarfs that from the ancient world for women's literacy."

136   The noblewoman Dhuoda: Contreni, "Learning for God."

136   "The *Physica*, consisting" / "monasteries became centres": Fiona Maddocks, *Hildegard of Bingen* (Faber and Faber, 2002), 150, 156.

137   This growing trend: Miles, "Origins and Development."

137   Artists began including: Miles, "Origins and Development." See also Erik Kwakkel, *Books Before Print* (Amsterdam University Press, 2018), 93–97.

137   "medieval best sellers": Wendy A. Stein, "The Book of Hours," Heilbrunn Timeline of Art History, Metropolitan Museum of Art, June 2017.

137 "Any Book of Hours" / "The female user": Eamon Duffy, *Marking the Hours* (Yale University Press, 2011), 36–38.
138 "Rather than simply mirroring": Pamela Sheingorn, "'The Wise Mother': The Image of St. Anne Teaching the Virgin Mary," *Gesta* 32, no. 1 (1993).
138 Rapidly multiplying monastic libraries: Matthew Battles, *Palimpsest* (Norton, 2015), 109.
138 "hack their way through": Vallejo, *Papyrus*, 336.

Marginalia: One Tongue to Another

139 Aristotle and al-Ma'mūn / "This dream": Uwe Vagelpohl and Ignacio Sánchez, "Why Do We Translate? Arabic Sources on Translation," in *Why Translate Science?*, ed. Dimitri Gutas (Brill, 2022), 275.
140 Gerard of Cremona: Violet Moller, *The Map of Knowledge* (Doubleday, 2019), 121–150; Falk, *Light Ages*, 86–87; David C. Lindberg, *The Beginnings of Western Science*, 2nd ed. (University of Chicago Press, 2007), 216–217.
140 "If I had control over": Falk, *Light Ages*, 111, 110.

Chapter 9: Hit Refresh

141 "I go to a spring" / "I have a book": Machiavelli, Letter to Francesco Vettori, December 10, 1513, in *The Essential Writings of Machiavelli*, trans. Peter Constantine (Modern Library, 2017), 508–510.
141 Once labored nine months: Maurizio Viroli, *Niccolò's Smile*, trans. Antony Shugaar (Farrar, Straus & Giroux, 2000), 7.
141 Viroli says he was partial: Viroli, *Niccolò's Smile*, 9.
142 Machiavelli copied: Chauncey E. Finch, "Machiavelli's Copy of Lucretius," *Classical Journal* 56, no. 1 (October 1960).
142 "When evening comes": Machiavelli, Letter to Vettori.
142 "I have noted down": Machiavelli, Letter to Vettori.
144 Scribes in one French monastery: G. Billanovich, "Petrarch and the Textual Tradition of Livy," *Journal of the Warburg and Courtauld Institutes* 14, no. 3/4 (1951).
144 Petrarch possessed a copy: Billanovich, "Petrarch and the Textual Tradition of Livy."
144 Around a century after Petrarch: Holt N. Parker, from his introduction to Antonio Beccadelli, *The Hermaphrodite*, ed. and trans. Holt N. Parker (Harvard University Press, 2010), xx.
145 Movement we know as Renaissance humanism: Jeffrey M. Hunt, R. Alden Smith, and Fabio Stok, *Classics from Papyrus to the Internet* (University of Texas Press, 2017), 150.
145 Petrarch woke many ancient manuscripts: Hunt, Smith, and Stok, *Classics from Papyrus*, 153, 155, 160–161.
145 "I want to know who," said Beccadelli: Parker, introduction to Beccadelli, *The Hermaphrodite*, xxxix.

145   "torn and mangled" / "Our descendants" / "to all mankind" / "public good": Ross King, *The Bookseller of Florence* (Atlantic Monthly Press, 2021), 25, 38.

146   Recirculating ancient wisdom: King, *Bookseller of Florence*, 38.

146   Petrarch tried learning: Colin Wells, *Sailing from Byzantium* (Delacorte, 2006), 53.

146   "Your Homer is here": Hunt, Smith, and Stok, *Classics from Papyrus*, 157.

146   Plato's *Republic* finally sound: Hunt, Smith, and Stok, *Classics from Papyrus*, 172–173.

147   "How ill I took it": Lisa Jardine, *Worldly Goods* (Norton, 1999), 57–58.

147   "I assembled almost all": Jardine, *Worldly Goods*, 62.

147   He eventually bequeathed them: Jardine, *Worldly Goods*, 63.

147   Not everyone involved in this movement: Hunt, Smith, and Stok, *Classics from Papyrus*, 176.

149   165 different translators: Hunt, Smith, and Stok, *Classics from Papyrus*, 174–175.

150   "The work would be": Michael Massing, *Fatal Discord* (Harper, 2018), 89–90.

150   "My readings in Greek": Massing, *Fatal Discord*, 88.

150   "Latin learning" / "a real Greek": Erasmus, Letter to Antony of Bergen, trans. Barbara Flower, in Johan Huizinga, *Erasmus and the Age of Reformation* (Harper, 1957), 202, 204.

150   Whenever he had money: Massing, *Fatal Discord*, 88.

150   More than a million copies: Andrew Pettegree, *The Book in the Renaissance* (Yale University Press, 2010), 84–85.

152   In his *Tractionum Theologicarum*: See table 2.2 in Jeffrey Mallinson, *Faith, Reason, and Revelation in Theodore Beza* (Oxford University Press, 2003), 40–41.

152   "My *Cicero* and *Plinies*": Thomas Heywood, *Pleasant Dialogues and Dramma's*, ed. W. Bang (David Nutt, 1903), 267. Per the earlier point about words becoming unfamiliar over time, some of these terms might vex. "Cothurnat" means singular. "Phaleucik" is a puzzler; it may refer to poetic meter.

154   "Google on parchment": Duncan, *Index*, 75.

154   "Hugh . . . will be the first" / "over 10,000 terms" / "wormholes": Duncan, *Index*, 55, 77, 61.

Marginalia: Paper Pushers

155   "I would rather be": Mark Kurlansky, *Paper* (Norton, 2016), 49.

156   Within several hundred years: Alexander Monro, *The Paper Trail* (Knopf, 2016), 214.

156   "In 1231 . . . the Holy Roman": Monro, *Paper Trail*, 222.

156   The thirteenth and fourteenth centuries: Monro, *Paper Trail*, 223; Nichole Howard, *The Book* (Johns Hopkins University Press, 2009), 10; Kurlansky, *Paper*, 97.

Chapter 10: Too Many Books?

157   "I am over the entrance": Montaigne, *Essays* 3.3 in *The Complete Works*, trans. Donald M. Frame (Everyman's Library, 2003).

157   "It was the most useless" / "In my library" / "There I leaf": Montaigne, *Essays* 3.3. In John Florio's Elizabethan translation, Montaigne "ransack[s]" his books.

157   "One moment I muse": Montaigne, *Essays* 3.3. I swapped "write" for "set" for clarity's sake.

157   Montaigne claimed he didn't: Montaigne, *Essays* 2.17.

158   He disavowed any facility: Montaigne, *Essays* 2.17, 3.3.

158   "I have adopted": Montaigne, *Essays* 2.10.

160   J. O. Ward worked out: J. O. Ward, "Alexandria and Its Medieval Legacy," in *The Library of Alexandria*, ed. Roy MacLeod (I. B. Tauris, 2004), 163–164.

160   The Pachomian project: Ward, "Alexandria and Its Medieval Legacy," 171.

160   Produced only around 120 books: Eltjo Buringh and Jan Luiten van Zanden, "Charting the 'Rise of the West': Manuscript and Printed Books in Europe, a Long-Term Perspective from the Sixth through Eighteenth Centuries," *Journal of Economic History* 69, no. 2 (June 2009). See also Eltjo Buringh, *Medieval Manuscript Production in the Latin West* (Brill, 2011), esp. chap. 5.

160   Most in Anglo-Saxon Britain: Michael Lapidge, *The Anglo-Saxon Library* (Oxford University Press, 2006), 57.

160   Evidence for monastic libraries: Lapidge, *Anglo-Saxon Library*, 58–59.

161   "By the end of this period": Ward, "Alexandria and Its Medieval Legacy," 171. University libraries, just beginning in this period, were puny. "It was not at all unusual for a college library at this time to have less than one hundred books," says philologist and librarian Ruth French Strout in "The Development of the Catalog and Cataloging Codes," *Library Quarterly* 26, no. 4 (1956).

161   Numbers for the Latin West: Buringh and van Zanden, "Charting the 'Rise of the West,'" table 1.

161   More than twelve million books: Buringh and van Zanden, "Charting the 'Rise of the West,'" table 2. Their estimates are consistent with that offered by Frédéric Barbier in *Gutenberg's Europe* (Polity, 2017), 239. Lucien Febvre and Henri-Jean Martin allow for as many as twelve million to twenty million; see *The Coming of the Book* (Verso, 2010), 248, 350. Andrew Pettegree and Arthur der Weduwen come in at just nine million; see *The Library*, 81.

161   They pressed out: Buringh and van Zanden, "Charting the 'Rise of the West,'" table 2; Barbier, *Gutenberg's Europe*, 239.

161   Three and four hundred thousand individual works: Barbier, *Gutenberg's Europe*, 239.

161   "By this art": Ann Blair, "The 2016 Josephine Waters Bennett Lecture: Humanism and Printing in the Work of Conrad Gessner," *Renaissance Quarterly* 70, no. 1 (2017).

161   Frédéric Barbier characterized: Barbier, *Gutenberg's Europe*, 239.

163   "the user must look up": José María Pérez Fernández and Edward Wilson-Lee, *Hernando Colón's New World of Books* (Yale University Press, 2021), 218.

164 "sixteen chests / Full of books": Pérez Fernández and Wilson-Lee, *Colón's New World*, 32–33.

165 Description of the *Epítomes*: Pérez Fernández and Wilson-Lee, *Colón's New World*, 43, 148–149, 154–155. See also Edward Wilson-Lee, *The Catalogue of Shipwrecked Books* (Scribner, 2019), 253–256.

166 Description of the *Materias*: Pérez Fernández and Wilson-Lee, *Colón's New World*, 45, 155–157, 164. See also Wilson-Lee, *Catalogue*, 268.

166 "The *Libro de las Materias* created a network": Pérez Fernández and Wilson-Lee, *Colón's New World*, 45.

167 "I rejoice and give thanks": Blair, "The 2016 Josephine Waters Bennett Lecture."

167 Together the books would give: Strout, "The Development of the Catalog."

167 "Although the typographical": Blair, "The 2016 Josephine Waters Bennett Lecture."

168 "Bodley envisioned": Pettegree and Weduwen, *The Library*, 146.

168 "If that indeed happened": Eric Berkowitz, *Dangerous Ideas* (Beacon, 2021), 25.

169 Anti-Catholic German peasants: Pettegree and Weduwen, *The Library*, 104–105.

169 Collectors valued some: Pettegree and Weduwen, *The Library*, 109–110.

170 Henry VIII published: Robin Vose, *The Index of Prohibited Books* (Reaktion, 2022), 49.

170 "Each edition's contents": Vose, *Index*, 50.

171 Bodley's successor at Oxford: Vose, *Index*, 87–88.

171 Section of a 1541 copy: Owen Jarus, "16th-Century Big Thinker Erasmus Was Censored," *NBC News*, May 15, 2012. See also Owen Jarus, "In Photos: A Tale of Two Censors," *Live Science*, May 15, 2012.

171 they "corrected" . . . *The Decameron*: Christian Algar, "Il Decamerone—'Corrected' by Rome," *European Studies Blog*, British Library, September 26, 2016.

171 Venetian book smugglers: Vose, *Index*, 109.

Marginalia: Readers Unleashed

173 "In 1485 Sultan Bayezid": Calestous Juma, *Innovation and Its Enemies* (Oxford University Press, 2016), 68.

173 "The great increase": Juma, *Innovation*, 75.

Chapter 11: Reading the Book of Nature

179 Gardner's comments on Darwin's notebooks: Rebecca Jones, "'Stolen' Charles Darwin Notebooks Left on Library Floor in Pink Gift Bag," BBC, April 4, 2022. See also Rebecca Jones, "Charles Darwin: Notebooks Worth Millions Lost for 20 Years," BBC, November 23, 2020.

180 "He wrote the notebook": David Quammen, *The Reluctant Mr. Darwin* (Norton, 2007), 28.

180 Newton meandered head down: Peter Ackroyd, *Newton* (Doubleday, 2006), 78.

180 "Thousands of sheets": James Gleick, *Isaac Newton* (Vintage, 2003), 124.

182   Anaximander of Miletus: Carlo Rovelli, *Anaximander*, trans. Marion Lignana Rosenberg (Riverhead, 2023). See also Maria Michela Sassi, *The Beginnings of Philosophy in Greece*, trans. Michele Asuni (Princeton University Press, 2018), 37–44.

183   "The important thing isn't": Colin Wells, "From Memory to Innovation," *Hedgehog Review*, Fall 2018.

183   First philosopher to write in prose: Sassi, *Beginnings of Philosophy*, 82.

183   "The invention of writing": Lindberg, *Beginnings of Western Science*, 11.

184   "beginning . . . of computational astronomy" / "As a result": Lindberg, *Beginnings of Western Science*, 16. Regarding Ptolemy's use of Babylonian data, see Rovelli, *Anaximander*, 7.

185   "arranged and exemplified": Benjamin Wardhaugh, *Encounters with Euclid* (Princeton University Press, 2021), 14.

186   John of Sacrobosco: Falk, *Light Ages*, 87.

186   "In the world of print": Wardhaugh, *Encounters with Euclid*, 77; for the printing statistics, see 66, 73, and 82. See also Margaret Bingham Stillwell, *The Awakening Interest in Science during the First Century of Printing* (Bibliographic Society of America, 1970), 50.

187   Burns's list of effects: I imposed the enumeration on William E. Burns, *The Scientific Revolution in Global Perspective* (Oxford University Press, 2016), 33–34.

188   Grueling collection of data: A point made by Michael Strevens in *The Knowledge Machine* (Liveright, 2020).

188   Edition of the Alfonsine Tables: Stillwell, *Awakening Interest in Science*, 8.

188   Vesalius's 1543 treatise: Burns, *Scientific Revolution*, 34; Stillwell, *Awakening Interest in Science*, 227–228; Febvre and Martin, *The Coming of the Book*, 277.

189   Ratdolt printed an edition: Stillwell, *Awakening Interest in Science*, 23.

189   Frankfurt printer Christian Egenolff: William Eamon, *Science and the Secrets of Nature* (Princeton University Press, 1994), 114–116.

189   Topsell's "monumental work": John H. Lienhard, "Episode No. 1586: Topsell's Beasts," *The Engines of Our Ingenuity*, n.d.

192   "I have included . . . 20,000 topics": Pliny, *Natural History*, preface in Vol. 1, trans. John Bostock and H. T. Riley (George Bell and Sons, 1893).

192   "For about thirty-five of them": Leroi, *The Lagoon*, 59.

193   "the elephant, the armadillo" / "fundamentally bookish": Sachiko Kusukawa, "Konrad Gessner," in *The Great Naturalists*, ed. Robert Huxley (Thames & Hudson, 2019), 66–67.

193   Dioscorides . . . covered five hundred plants: James Poskett, *Horizons* (Mariner, 2022), 25.

194   "At the beginning of the seventeenth century": Poskett, *Horizons*, 138.

194   "the Nigroæ, whose king": Pliny, *Natural History* 6.35, LCL 370 (Ethiopia).

194   "We are not bound to believe": Augustine, *The City of God* 16.8, in *Nicene and Post-Nicene Fathers*, Vol. 2, first series, ed. Philip Schaff (Hendrickson, 1994).

194   "heads like dogs": Marco Polo, *The Travels of Marco Polo*, trans. Henry Yule and Henri Cordier (Dover, 1993), 2:309.

194   "I have not found": Christopher Columbus, *The Four Voyages of Christopher Columbus*, trans. J. M. Cohen (Penguin, 1969), 121.

194   So-called torrid zone: Burns, *Scientific Revolution*, 39.

195   "The solution to the sterility": Burns, *Scientific Revolution*, 36.

195   Revival of naturalistic explanations: Lindberg, *Beginnings of Western Science*, 364–365.

196   William Burns's list: Burns, *Scientific Revolution*, 33.

196   "They [also] had the printing press": David Wootton, *The Invention of Science* (Harper, 2015), 51.

196   Nine hundred individual first editions: See Stillwell, *Awakening Interest in Science*.

197   Astronomer Tycho Brahe: Burns, *Scientific Revolution*, 33.

198   "Darwin . . . relied on mathematical" / "Though he was not": Vassiliki Betty Smocovitis, "Darwin's Botany in the *Origin of Species*," in *The Cambridge Companion to the "Origin of Species"*, ed. Michael Ruse and Robert J. Richards (Cambridge University Press, 2009), 221–222.

198   Wallace had also read: Janet Browne, *Darwin's* Origin of Species (Atlantic Monthly Press, 2006), 61–62.

Marginalia: The Third Variable

199   The stasist/dynamist distinction: See Virginia Postrel, *The Future and Its Enemies* (Free Press, 1998).

Chapter 12: Founded on Books

201   In the summer of 1775: Thomas Kidd, *Thomas Jefferson: A Biography of Spirit and Flesh* (Yale University Press, 2022), 45–46.

201   "Sensible that I labour": Thomas Jefferson, letter to Lucy Ludwell Paradise, June 1, 1789, Monticello.org.

201   He bummed $133: Kidd, *Thomas Jefferson*, 80.

201   "I am now entered on": Thomas Jefferson, letter to Thomas Law, April 23, 1811, Monticello.org.

203   "Americans were . . . remarkably literate": Forrest McDonald, "A Founding Father's Library," *Literature of Liberty* 1, no. 1 (January/March 1978).

203   By 1790, more than sixty different: Charles E. Clark, "Early American Journalism," in *A History of the Book in America*, Vol. 1: *The Colonial Book in the Atlantic World*, ed. Hugh Amory and David D. Hall (University of North Carolina Press, 2007), 361.

203   "hurtful" / "corrupting" / "atmosphere of falsehoods": James Fenimore Cooper, *The American Democrat* (Liberty Fund, 1981), 160–161.

204   "Every lover of his country": McDonald, "A Founding Father's Library."

204   By 1771 . . . Boston boasted: Hugh Amory, "The New England Book Trade, 1713–1790," in Amory and Hall, *A History of the Book in America: The Colonial Book in the Atlantic World*, 1:332.

204   Book stocks were meager: Robert C. Baron and Conrad Edick Wright, *The Libraries, Leadership, and Legacies of John Adams and Thomas Jefferson* (Fulcrum, 2010), xvi–xvii.

204   "I remember reading" / "shelf of rough-hewn": Alexis de Tocqueville, *Democracy in America*, trans. Arthur Goldhammer (Library of America, 2004), 538, 864.

205   "Charles Rollin's two-volume *The Ancient History*": McDonald, "A Founding Father's Library."

205   "Our laws, language, religion": Thomas Jefferson, letter to William Duane, August 12, 1810, in *Writings*, ed. Merrill D. Peterson (Library of America, 1984), 1228.

205   "proper for the use": "Notes on Debate," January 23, 1783, *Founders Online*, National Archives.

206   His finished list: "Report on Books for Congress," January 23, 1783, *Founders Online*, National Archives.

206   "In the purchase of books": Thomas Jefferson, letter to James Madison, May 8, 1784, *Founders Online*, National Archives.

206   "I shall subjoin" / "I have been induced": Thomas Jefferson, letter to James Madison, November 11, 1784, *Founders Online*, National Archives.

207   "I mentioned to you": Thomas Jefferson, letter to James Madison, March 18, 1785, *Founders Online*, National Archives.

207   "I thank you much" / "I am afraid" / "treatises on the antient" / "where they can": James Madison, letter to Thomas Jefferson, April 27, 1785, *Founders Online*, National Archives.

209   Jefferson was prevented: Thomas Jefferson, letter to James Madison, May 11, 1785, *Founders Online*, National Archives.

209   "I have not yet received": James Madison, letter to Thomas Jefferson, August 20, 1785, *Founders Online*, National Archives.

209   "I have at length made": Thomas Jefferson, letter to James Madison, September 1, 1785, *Founders Online*, National Archives.

209   "Upon returning to America": Mark Dimunation, "The Whole of Recorded Knowledge," in Baron and Wright, *The Libraries, Leadership, and Legacies of John Adams and Thomas Jefferson*, 28.

209   "Since I have been at home": James Madison, letter to Thomas Jefferson, March 18, 1786, *Founders Online*, National Archives.

210   "It is clear that the delay": James Madison, "Notes on Ancient and Modern Confederacies, [April–June?] 1786," *Founders Online*, National Archives. The brackets around 1584 exist in Madison's original.

211   "Every Person seems" / "He blends together": William Pierce, "Notes of Major William Pierce (Georgia) in the Federal Convention of 1787," Avalon Project, Yale Law School.

212   On April 18, Washington asked: George Washington, letter to the cabinet, April 18, 1793, *Founders Online*, National Archives. See also "Editorial Note: Jefferson's Opinion on the Treaties with France," *Founders Online*, National Archives.

212   "the people [are] the source": "Notes on Washington's Questions on Neutrality and the Alliance with France," undated, *Founders Online*, National Archive.

212   "the nation's most learned": McDonald, "A Founding Father's Library."

212   Jefferson had a copy of Vattel: Jefferson had been reading Vattel for years, as is clear from passages in the Declaration of Independence and his autobiography. See William Ossipow and Dominik Gerber, "The Reception of Vattel's Law of Nations in the American Colonies," *American Journal of Legal History* 57, no. 4 (2017); Jefferson, *Autobiography*, in Peterson, *Writings*, 51 (Jefferson's footnote on ratifying treaties directly cites Vattel).

213   "Would you suppose": Thomas Jefferson, letter to James Madison, April 28, 1793, *Founders Online*, National Archives.

213   "the validity of treaties": "Notes on Washington's Questions."

213   He continued thinking on paper: "Notes for Opinion on the Treaty of Alliance with France," undated, *Founders Online*, National Archives.

214   "for the purchase of such books": Act of April 24, 1800. Chap. 38, § 5. 6 Stat. 56. See also Ainsworth R. Spofford, "Our Leading Libraries," *Magazine of American History* 29, no. 1 (January 1893).

215   "I learn from the Newspapers" / "I presume" / "for such works" / "such a collection" / "hardly probable": Thomas Jefferson, letter to Samuel H. Smith, September 21, 1814, in Peterson, *Writings*, 1353.

216   Sale of Jefferson's library: "Editorial Note," *Founders Online*, National Archives, https://founders.archives.gov/documents/Jefferson/03-07-02-0484-0001; Kidd, *Thomas Jefferson*, 200–202; Dimunation, "The Whole of Recorded Knowledge," 22–23.

216   Washington . . . was more bookish: See Kevin J. Hayes, *George Washington: A Life in Books* (Oxford University Press, 2017).

216   Adams carried copy of Cicero: R. B. Bernstein, "Let Us Dare to Read, Think, Speak, and Write," in Baron and Wright, *The Libraries, Leadership, and Legacies of John Adams and Thomas Jefferson*, 85, 87. For Adams's book collecting, see Beth Prindle, "Thought, Care, and Money," in the same volume.

216   "a common matrix": McDonald, "A Founding Father's Library." See also Bernstein, "Let Us Dare to Read," 85.

217   "a canine appetite": Thomas Jefferson, letter to John Adams, May 17, 1818, Monticello.org.

Marginalia: Uncommon Impact

218   Days leading up to independence: Bernard Bailyn, *The Ideological Origins of the American Revolution*, enl. ed. (Belknap, 1992), x.

218   "The aim of almost every . . . notable pamphlet": Bernard Bailyn, *Faces of Revolution* (Vintage, 1992), 82.

219   "One of the vilest things": Henry Mayer, *A Son of Thunder* (Grove Press, 1991), 287.

219 "A most flagitious Performance": Ambrose Serle's diary entries in *The American Revolution*, ed. John Rhodehamel (Library of America, 2001), 149.

219 "Poor, ignorant, malicious" / "Paine's aim . . . was to tear": Bailyn, *Faces*, 71, 82.

220 Popular estimates range: Trish Loughran, *The Republic in Print* (Columbia University Press, 2007), chap. 2.

220 "Everyone read it": Jill Lepore, *Book of Ages* (Vintage, 2013), 181.

220 *Common Sense* didn't cause: Bailyn, *Faces*, 84.

Chapter 13: Literature for Liberation

221 On the Hartford flood and Pennington's book: Christopher L. Webber, *American to the Backbone* (Pegasus, 2011), 148–150.

223 "When first I saw" / "I follow'd him": James Albert Ukawsaw Gronniosaw, *A Narrative of the Most Remarkable Particulars in the Life of James Albert Ukawsaw Gronniosaw, an African Prince, as Related by Himself*, in *Slave Narratives*, ed. William L. Andrews and Henry Louis Gates Jr. (Library of America, 2000), 11–12.

224 "I had a great curiosity": Olaudah Equiano, *The Interesting Narrative of the Life of Olaudah Equiano*, in Andrews and Gates, *Slave Narratives*, 86.

224 Gronniosaw was at first "uneasy": Gronniosaw, *Narrative*, 14.

224 More than a hundred pupils: David Hackett Fischer, *African Founders* (Simon & Schuster, 2022), 168.

225 "endeavoring to make letters": David Waldstreicher, *The Odyssey of Phillis Wheatley* (Farrar, Straus & Giroux, 2023), 3.

225 "Did Fear and Danger": Carl Bridenbaugh and Phillis Wheatley, "The Earliest-Published Poem of Phillis Wheatley," *New England Quarterly* 42, no. 4 (1969), 583–584.

225 "Thou didst in strains": Phillis Wheatley, *Poems on Various Subjects Religious and Moral* (W. H. Lawrence & Co., 1887), 21.

226 "We are . . . a race of Beings": Benjamin Banneker, letter to Thomas Jefferson, August 19, 1791, *Founders Online*, National Archives.

227 "She has been examined" / "uncultivated" / "a slave": Wheatley, *Poems*, 9.

227 "Never yet could I" / "Misery is often" / "their slaves were often" / "Epictetus, Diogenes, Phaedon": Thomas Jefferson, *Notes on the State of Virginia*, Query XIV, in Peterson, *Writings*, 266–268.

228 "Wheatley takes up": Waldstreicher, *Odyssey of Phillis Wheatley*, 307.

228 As one college president: Waldstreicher, *Odyssey of Phillis Wheatley*, 344.

229 "We wish to plead": David Shedden, "Today in Media History: The First African-American Newspaper, Freedom's Journal, Was Founded in 1827," Poynter Institute, March 16, 2015.

229 "For colored people to acquire": David Walker, *Appeal to the Coloured Citizens of the World*, 3rd ed. (David Walker, 1830), 37.

230 For Walker's treatise in Georgia and the state's response: Heather Andrea Williams, *Self-Taught* (University of North Carolina Press, 2005), 14–15, 203–204;

E. Jennifer Monaghan, "Reading for the Enslaved, Writing for the Free," *Proceedings of the American Antiquarian Society* 108, pt. 2 (October 1998).

230 For Louisiana's response: Williams, *Self-Taught*, 204–205.

230 Stono River outside Charleston: Mohammed Elnaiem, "Did Kongolese Catholicism Lead to Slave Revolutions?" *JSTOR Daily*, February 6, 2019; Williams, *Self-Taught*, 13.

231 South Carolina's 1740 Slave Code: Willie Lee Rose, *Rehearsal for Reconstruction* (Bobbs-Merrill, 1964), 86.

231 Freedman Denmark Vesey: Jeremy Schipper, *Denmark Vesey's Bible* (Princeton University Press, 2022).

231 Nat Turner's 1831 revolt: Anthony Santoro, "The Prophet in His Own Words: Nat Turner's Biblical Construction," *Virginia Magazine of History and Biography* 116, no. 2 (2008).

231 Jacobs's home being searched: Harriet Jacobs, *Incidents in the Life of a Slave Girl* (Broadview Editions, 2023), chap. 12.

233 A wave of laws: Williams, *Self-Taught*, 203–213; Monaghan, "Reading for the Enslaved."

233 "Laws have been recently passed": James Henry Hammond, *Gov. Hammond's Letters on Southern Slavery* (Walker & Burke, 1845), 11–12.

233 "While reading this speech": John Thompson, *The Life of John Thompson, a Fugitive Slave* (John Thompson, 1856), 38.

234 Literacy among runaways: Fischer, *African Founders*, 169, 343.

234 "ke' dat up deir sleeve": Janet Cornelius, "'We Slipped and Learned to Read': Slave Accounts of the Literacy Process, 1830–1865," *Phylon* 44, no. 3 (1983). Using a database of 3,428 former slaves interviewed by the Federal Writers' Project, Janet Cornelius calculated just over 5 percent had learned to read and write as slaves, though she points out the number is possibly higher: "Neither slaves nor those slaveowners and other whites who taught them could proclaim their activities safely."

234 "privilege, which so rarely falls": Jacobs, *Incidents*, 76.

234 Testimonies left by 272 literate former slaves: Cornelius, "'We Slipped and Learned to Read.'"

234 "Bible could be a double-edged": See Allen Dwight Callahan, *The Talking Book* (Yale University Press, 2006).

234 "fugitive learners": See Jarvis R. Givens, *School Clothes* (Beacon, 2023), e.g., chap. 2.

234 Taylor's story: Susie King Taylor, *Reminiscences of My Life in Camp* (Susie King Taylor, 1902), 5–6.

236 For the locations of fugitive learners' schools: See Givens, *School Clothes*, chap. 2.

236 Frederick Douglass helped out: Frederick Douglass, *Autobiographies*, ed. Henry Louis Gates Jr. (Library of America, 1994), 559. These quotes are taken from *The Life and Times of Frederick Douglass*, chap. 14. He tells the same story in his two prior memoirs as well; in the first, he refers to Wilson as White.

236 "Slaves . . . caught writing": Callahan, *The Talking Book*, 9.

236 "The first time you was caught" / "had the white doctor": Cornelius, "'We Slipped and Learned to Read.'"

237 Two slaves . . . practiced: Fischer, *African Founders*, 346.

237 Neighborhood kids taught William Anderson: Fischer, *African Founders*, 345.

237 In the nursery: Fischer, *African Founders*, 344–345; Cornelius, "'We Slipped and Learned to Read.'"

237 He swapped bread: Douglass, *Autobiographies*, 223–225. Here I quote from the second of his memoirs, *My Bondage and My Freedom*. He tells the same story in his first, *Narrative of the Life of Frederick Douglass*, 41, in the Library of America edition.

238 Douglass encountering the two orations in *The Columbian Orator*: Douglass, *Autobiographies*, 41–42.

239 For Pennington's story: See James W. C. Pennington, *The Fugitive Blacksmith*, 2nd ed. (Charles Gilpin, 1849); Webber, *American to the Backbone*.

240 "There is one sin": Pennington, *Fugitive Blacksmith*, 56–57.

Marginalia: Malcolm X's Alma Mater

242 For Malcolm X's story and all quotes from Malcolm X: Malcolm X and Alex Haley, *The Autobiography of Malcolm X* (Penguin, 2001).

242 "While you're in there": Wilfred Little, "Our Family from the Inside," *Contributions in Black Studies* 13/14 (1995/1996).

243 "I sit with Shakespeare": W. E. B. Du Bois, *The Souls of Black Folk* (Penguin, 2018), 84.

Chapter 14: Seeing with Other Eyes

245 "She has been to school": Charles Edward Stowe, *Life of Harriet Beecher Stowe Compiled from Her Letters and Journals* (Houghton Mifflin, 1890), 8.

245 "intelligent & studious": David S. Reynolds, *Mightier Than the Sword* (Norton, 2011), 5.

245 "This room had to me": Stowe, *Life of Harriet Beecher Stowe*, 9.

245 "There were Bell's Sermons": Stowe, *Life of Harriet Beecher Stowe*, 10.

246 "What wonderful stories": Stowe, *Life of Harriet Beecher Stowe*, 10.

246 Peeking through these castoffs: Stowe, *Life of Harriet Beecher Stowe*, 9; Martha Foote Crow, *Harriet Beecher Stowe* (D. Appleton and Company, 1913), 74; Reynolds, *Mightier Than the Sword*, 5.

246 "I have always said": Stowe, *Life of Harriet Beecher Stowe*, 25.

246 In one of her novels: Harriet Beecher Stowe, *The Minister's Wooing* (Sampson Low, Son, & Co., 1859), 17–18.

246 "Here I loved to retreat": Stowe, *Life of Harriet Beecher Stowe*, 9–10.

248 "With too little sleep" / "When his mind": Miguel de Cervantes, *Don Quixote*, trans. Edith Grossman (Ecco, 2005), 21.

249 "One day we read": Dante, *Inferno*, Canto 5.

250   "Our Galehot": Dante, *Inferno*, Canto 5.

250   "The pimp was the book": "Seth Lerer on the History of the Book," *Entitled Opinions with Robert Harrison*, April 3, 2006.

250   "torrents of emotion": Lynn Hunt, *Inventing Human Rights* (Norton, 2007), 36.

250   "In France . . . 8 new novels": Hunt, *Inventing Human Rights*, 40. For the British statistics, see Franco Moretti, "Style, Inc. Reflections on Seven Thousand Titles (British Novels, 1740–1850)," *Critical Inquiry* 36, no. 1 (Autumn 2009): 139–140.

250   275 novels in time: Reinhard Wittman, "Was There a Reading Revolution at the End of the Eighteenth Century?" in *A History of Reading in the West*, ed. Guglielmo Cavallo and Roger Chartier (University of Massachusetts Press, 2003), 304.

251   A single novel: Wittman, "Was There a Reading Revolution," 303.

251   "Everyone, but women": Wittman, "Was There a Reading Revolution," 285.

251   "Here people are stuffed": Wittman, "Was There a Reading Revolution," 298.

251   "Reading mania" / "No lover of tobacco": Wittman, "Was There a Reading Revolution," 300, 285.

251   "It takes possession, all night": Hunt, *Inventing Human Rights*, 45.

252   "Strokes penetrate immediately" / "I never felt so much" / "My Spirits are strangely seized": Hunt, *Inventing Human Rights*, 49, 46, 46.

253   "I have felt pass" / "You have driven": Hunt, *Inventing Human Rights*, 48.

253   "Why . . . can't we investigate": Martha C. Nussbaum, *Love's Knowledge* (Oxford University Press, 1990), 47.

254   "Reading is . . . a technology": Steven Pinker, *The Better Angels of Our Nature* (Penguin, 2011), 175.

254   "Our experience is, without fiction": Nussbaum, *Love's Knowledge*, 47–48.

254   "intellectual sex": "Writing Fiction, Overcoming Depression, and Ben Shapiro Becoming Christian (Andrew Klavan)," *Pints with Aquinas Podcast*, October 16, 2024.

254   "We demand windows" / "every act of justice": C. S. Lewis, *An Experiment in Criticism* (Cambridge University Press, 2023), 137–138.

255   By loaning our own cognitive: Keith Oatley, *Such Stuff as Dreams: The Psychology of Fiction* (Wiley-Blackwell, 2011), 116–117.

255   Cathartic expression: Oatley, *Such Stuff as Dreams*, 123.

255   "The emotions we experience": Oatley, *Such Stuff as Dreams*, 115.

255   "Art . . . enables us to experience": Oatley, *Such Stuff as Dreams*, 118.

256   All else being equal, research shows / "moral laboratory": Oatley, *Such Stuff as Dreams*, 159, 172.

256   "The magical spell cast": Hunt, *Inventing Human Rights*, 58.

257   "I remember distinctly" / "I will write something": Stowe, *Life of Harriet Beecher Stowe*, 145.

259   "We confess to the frequent": Reynolds, *Mightier Than the Sword*, 130.

259   "The touching, but too truthful" / "so stirred": Reynolds, *Mightier Than the Sword*, 129–130.

259   On the Southern reaction: Reynolds, *Mightier Than the Sword*, 150–167.

Marginalia: The Futility of Banning Books

261 Attempts at banning: "'The Color Purple,' a Pulitzer Prize–Winning Novel, Has Been . . . ," UPI, May 3, 1984.

261 Over the last thirty-odd: "100 Most Frequently Challenged Books by Decade," American Library Association, n.d.

261 In the mid-1980s: E. R. Shipp, "Blacks in Heated Debate over 'The Color Purple,'" *New York Times*, January 27, 1986.

261 "I opened the page" / "the single most" / "life would have": Salamishah Tillet, *In Search of* The Color Purple (Abrams Press, 2021), 130, 132, 129.

263 "Book bans inhibit": David French, "The Dangerous Lesson of Book Bans in Public School Libraries," *Reason*, August/September 2022.

Chapter 15: Browsing the Universal Library

265 "Thinking Machine": James Gleick, *The Information* (Vintage, 2012), 171.

265 "We are becoming bogged" / "we adhere rather closely": James M. Nyce and Paul Kahn, eds., *From Memex to Hypertext* (Academic Press, 1991), 42.

266 Inventor Richard Hoe: Scott E. Casper, "Introduction," in *A History of the Book in America*, Vol. 3: *The Industrial Book, 1840–1880*, ed. Scott E. Casper et al. (University of North Carolina Press, 2007), 1.

268 "It seemed a wonder": Alex Wright, *Cataloguing the World* (Oxford University Press, 2014), 47–48.

268 "The ideal . . . would be to strip": Wright, *Cataloguing the World*, 80.

269 Breaking free from simple hierarchies: Wright, *Cataloguing the World*, 85.

270 three million index cards / "an inventory of all": Wright, *Cataloguing the World*, 69, 76.

270 at least one acquaintance: Wright, *Cataloguing the World*, 218. As Wright notes, many others were working on the same problem.

270 "growing mountain of research" / "Mendel's concept" / "It involves the entire": Vannevar Bush, "As We May Think," *Atlantic Monthly*, July 1945, 101–102, 105.

271 "Long banks of shelves" / "the content of a thousand": Vannevar Bush, "The Inscrutable Thirties," in Nyce and Kahn, *From Memex to Hypertext*, 74.

272 "It is exactly as though": Bush, "As We May Think," 107.

272 Otlet had a similar idea: Alex Wright, *Informatica: Mastering Information through the Ages* (Cornell University Press, 2023), 158.

272 "Bush's great insight": G. Pascal Zachary, "The Godfather," *Wired*, November 1, 1997.

273 "Professional societies": Vannevar Bush, "Memex II," reprinted in Nyce and Kahn, *From Memex to Hypertext*, 172.

273 In 1963, he placed tongue: J. C. R. Licklider, "Intergalactic Network," reprinted in M. Mitchell Waldrop, *The Dream Machine* (Stripe Press, 2018), 526.

274 He argued as much: See "The Computer as a Communication Device," in Waldrop, *The Dream Machine*, 503–518.

274 "Often, the information" / "An intriguing possibility": Tim Berners-Lee, "Information Management: A Proposal," CERN, March 1989.

275 "Much needs to occur" / "For mature thought": Bush, "As We May Think," 104.

275 From time immemorial / "Now man takes a new step": Vannevar Bush, "Science Pauses," in Nyce and Kahn, *From Memex to Hypertext*, 189–190.

275 "it demonstrated . . . that machines": Belinda Barnet, "The Technical Evolution of Vannevar Bush's Memex," *DHQ: Digital Humanities Quarterly* 2, no. 1 (2008).

275 "Whenever logical processes": Bush, "As We May Think," 105.

276 As Bush's plans for the Memex: Vannevar Bush, *Science Is Not Enough* (William Morrow, 1967), 96. See also Vannevar Bush, *Pieces of the Action* (William Morrow, 1970), 191–192.

276 The printed page "is superb": J. C. R. Licklider, *Libraries of the Future* (MIT Press, 1965), 4.

276 "We may seek out inefficiencies": Licklider, *Libraries of the Future*, 5.

278 "to think in interaction": J. C. R. Licklider, "Man-Computer Symbiosis," reprinted in Waldrop, *The Dream Machine*, 492.

278 "procognitive systems": Licklider, *Libraries of the Future*, 6.

279 "searching, calculating, plotting, transforming" / "about 85 percent of my 'thinking' time": Licklider, "Man-Computer Symbiosis," 492–493.

279 "A basic part of the overall aim": Licklider, *Libraries of the Future*, 32.

Marginalia: Back to the Beginning

281 Fewer than a hundred people: Alison George, "How the Secrets of Ancient Cuneiform Texts Are Being Revealed by AI," *New Scientist*, August 3, 2022.

281 Using five thousand images: Edward C. Williams et al., "DeepScribe: Localization and Classification of Elamite Cuneiform Signs via Deep Learning," *arXiv* (2023).

281 Another team of researchers: Gai Gutherz et al., "Translating Akkadian to English with Neural Machine Translation," *PNAS Nexus* 2, no. 5 (2023).

281 "Translating all the tablets": Quoted in Melanie Lidman, "Groundbreaking AI Project Translates 5,000-Year-Old Cuneiform at Push of a Button," *Times of Israel*, June 17, 2023.

282 "There is all sorts of information": Jessica Sieff, "Researchers Use AI to Unlock the Secrets of Ancient Texts," University of Notre Dame, College of Engineering, August 5, 2021.

283 By early 2024: Taylor Nicioli, "Researchers Reveal First Full Passages Decoded from Famously Inscrutable Herculaneum Scrolls," CNN, February 7, 2024.

Chapter 16: Engines of Change

287   "I would never read a book": Adam Fisher, "Sam Bankman-Fried Has a Savior Complex—and Maybe You Should Too," Sequoia Capital, September 22, 2022. After the FTX collapse, Sequoia yanked the article. The inquisitive can still find it at the Internet Archive's Wayback Machine and Archive.Today.

287   "It's impossible to read": David Streitfeld, "Sam Bankman-Fried's Wild Rise and Abrupt Crash," *New York Times*, November 2, 2023.

288   Manual for Civilization: Ahmed Kabil, "How Can We Create a Manual for Civilization?" LongNow.org, June 7, 2017.

289   "People like to think": Douglas Rushkoff, "Technologies Have Biases," in *This Will Make You Smarter*, ed. John Brockman (Harper, 2012), 41–42.

292   "Without books, history is silent": Barbara Tuchman, "Papyrus to Paperbacks," *Washington Post*, December 30, 1979.

292   "A book is . . . a being": Epigraph nicked from Lina Bolzani, *A Marvelous Solitude* (Harvard University Press, 2023), 1.

INDEX

www.ingramcontent.com/pod-product-compliance
Lightning Source LLC
Chambersburg PA
CBHW051729260326
41914CB00040B/2025/J